Vintage
Ford
Tractors

Text by
Robert N. Pripps
Photographs by
Andrew Morland

RAINCOAST BOOKS
Vancouver

Dedication

For my latest grandson, Marcus Steven Pripps
—Robert N. Pripps

Text copyright © 1997 by Robert N. Pripps
Photographs copyright © 1997 by Andrew Morland

Edited by Michael Dregni
Designed by Andrea Rud
Printed in Hong Kong

97 98 99 00 01 5 4 3 2

Canadian Cataloguing in Publication Data
Pripps, Robert N., 1932–
    Vintage Ford tractors
    Includes bibliographical references and index.
    ISBN 1-55192-099-9
    1. Ford tractors—History. 2. Ford tractors—Pictorial works. I. Morland, Andrew. II. Title.
    TL233.6.F66P74 1997   629.255'2   C97-910308-8

First published in Canada in 1997 by
Raincoast Books
8680 Cambie Street
Vancouver, B.C. V6P 6M9
(604) 323-7100

First published in the United States in 1997 as a Town Square Book by
Voyageur Press, Inc., Stillwater, MN

Page 1: *A lineup of classic Fords preceded by a 1925 Fordson F with an Oliver Fordson plow. Owner: Don Artman.*
Page 2–3: *1950s Funk-Ford V-8 8N. Owner: Palmer Fossum.*
Page 3, inset: *A Fordson with the 1926 Hadfield-Penfield Model X Rigid Rail Tracks outfit.*
Facing page: *1955 Ferguson TED-20. Owner: Brian Whitlock.*

Contents

Acknowledgments

Andrew Morland and I extend our heartfelt thanks to all who washed, gassed, and moved tractors, charged batteries, mowed lawns, fixed lunches, invited friends, and prayed for sunshine.

Thanks to Lanette K. Clumill, Research Center, Henry Ford Museum & Greenfield Village, for working with me in the selection of archive photos.

Special thanks to Tom Armstrong and his wife, Cindy, of N-Complete, who put on a two-day display in Wilkinson, Indiana, inviting more than a dozen of their remanufactured-tractor customers back with their tractors for us to photograph.

We are most grateful to the Ford Motor Company: Jim Trainor and Scott Jensen of the Chicago Public Affairs Office for the new Ford Taurus Station Wagon that hauled us and our equipment on this photographic junket. The car was arranged through Ms. Murial Novi of the U.S. Auto Movers, Des Plaines, Illinois.

Thanks to Dr. Les Stegh, Archivist of Deere & Company, who graciously provided photos of Ford tractors with Deere equipment.

Thanks to Kelly Jewell of Tractor Implement Supply Company, Nashville, Tennessee, and a Ford tractor collector, for loads of information on the old Fords.

Just talking on the phone with eighty-six-year-old Joe Funk was a kick. Then Joe, the surviving Funk twin brother, sent photos, books, videos, and audio tapes that were invaluable in writing the chapter on Funk-Ford conversions. Thanks so much, Joe, and best regards.

Thanks, and a tip of my Ford-Ferguson hat, to Dean Simmons, collector and restorer from Fredricktown, Ohio, for supplying me with reference material on English Fordsons.

Special thanks, and apologies, to Daniel Simecek, Twinsburgh, Ohio, who sent photos of his restored tractors when our time schedule prevented us from getting to his place.

The collectors we visited were all members of the Ford/Fordson Collectors Association. The roster of members included those with extensive and interesting tractor collections that were just outside the range of our time-restricted travels. Anyone with an interest in old Ford Tractors needs to belong to this association. Contact Jim Ferguson (no relation), 645 Loveland-Miamiville Road, Loveland, OH 45140.

Thanks as well to tractor enthusiasts in North America and Europe, including Don Artman, Dale Bissen, Fred Bissen, Eric Coates, Jack Crane, Darrell Craycraft, Dennis Crossman, Floyd Dominique, Ken and Margaret Ellis, Dwight Emstrom, Palmer Fossum, Dr. Larry George, Mike Hanna, Charles Hardesty, Ron Lamoloy, Doug Marcum, Charlene Meyer, Robert Meyer, Louis Norfleet, Jonathan Philip, Duke Potter, Gene Runkle, Carleton Sather, Jim Spark, Ivan Sparks, Ron and Shirley Stauffer, Brian Whitlock, Jim and Jane Woehrman, and to Phil Green of the Fordson 500 National Vintage Tractor and Engine Club Rally.

A special thanks to Mike Weaver and Gene Hemphill of New Holland North America, Inc.

A final thanks to Michael Dregni, Editorial Director of Voyageur Press, for inviting us to become part of the Voyageur Press family.

Robert N. Pripps
Andrew Morland

Fordson

Ample Power for Quick Economical Threshing

Fordson Power is never more appreciated than at harvest time when threshing must be handled on the most economical basis to insure satisfactory profits.

On thousands of farms this year Fordson Tractors will furnish the power for quick, thorough and economical threshing. Fewer men with Fordson Power will thresh more grain at a lower cost. Bigger grain profits will result.

And with the threshing done Fordson Power is ready to bale hay, grind feed, fill the silo, saw wood, handle your fall plowing, etc., all at a big saving in time and money. On farm jobs of every description and for all belt work, Fordson Power proves the most profitable farm investment you can make.

Now—before harvest—is the time to buy your Fordson. See your nearest authorized Ford dealer today.

Ford Motor Company

CARS · TRUCKS · TRACTORS

Detroit, Michigan

See the Nearest Authorized Ford Dealer

Foreword

by Harold L. Brock

Ford N Series tractor design engineer 1939–1958
Deere & Company New Generation tractor engineering executive 1959–1985
Society of Automotive Engineers president 1971

I worked directly with Henry Ford as the designer of the Ford N Series farm tractors. Looking back now as an octogenarian, I am proud to have been a participant in the development of many of the tractors covered in Robert N. Pripp's excellent book, *Vintage Ford Tractors.*

The job of designing the Ford Model 9N tractor was assigned to me in 1939 with Mr. Ford's approval. The 9N tractor project set an all-time record from concept to production. Our instructions were to design a tractor to replace the horse and mule, and not to pay any attention to what was already on the market. The tractor was to cost a buyer no more than a team of horses, their harnesses, and the ten acres of land necessary to feed the animals.

Upon receiving the project assignment, I quickly found out that I had many project leaders. Foremost was Mr. Ford, Sorensen, and my immediate boss, Chief Engineer Lawrence Sheldrick. In addition, I had Ferguson, Willie Sands, and John Chambers offering comments and suggestions on details mainly concerning the hitch and its control. Seldom could I get them together as a group, and I had great difficulty assuring them that my design group was following each of their suggestions.

The Ferguson group did no tractor design work. They were to be responsible for the implements used with the tractor. Upon completion of the tractor, we found that the implements Ferguson had to offer were unacceptable for use in the North American market, and we had to quickly design a new plow. Sorensen, having been a pattern maker before becoming Ford's key executive, suggested the plow be designed with cast-steel construction instead of the usual rerolled railroad rail, thus obtaining a lightweight product. We

Harold L. Brock today, displaying Harry Ferguson's spring-wound tractor model that showed the basic workings of the three-point hitch system. This model was first shown by Ferguson to Henry Ford at the meeting that formed the famous Handshake Agreement.

also had to design a cultivator and adapt available planters, disc harrows, and mowers to the three-point hitch.

With the rush to get into production, Mr. Ford had no idea of the cost of the tractor. Ferguson prevailed upon Mr. Ford that his objective of low cost should be accomplished. Ferguson showed Mr. Ford an Allis-Chalmers Model B tractor, which was the lowest-cost tractor on the market. The Allis-Chalmers tractor was a stripped-down model with handcrank starting, a drawbar, and little else. It sold for approximately $500. The 9N was a styled tractor, having a self starter, hydraulic hitch, and many other features. Mr. Ford agreed to market it for $585.

The famous Gentlemen's Agreement between Mr. Ford and Harry Ferguson was relayed to Ford Chief Executive Charles Sorensen as "Ford will design, develop, and produce the tractor and Ferguson will distribute it." Ford management was concerned about the interpretation Ferguson would place on such a broad agreement, yet Sorensen was reluctant to discuss the agreement with Mr. Ford. Ferguson, therefore, took advantage of the situation during his association with Ford Motor Company and was difficult at times to deal with.

Ferguson's well-orchestrated plan of obtaining a tractor that would bear his name was finally accomplished after the break with Ford. He had confided in me his desire for greater recognition on the nameplate. I suggested that without the Ford name the tractor would never be accepted.

Ford tractor production was eventually merged into the car production process, yet Ford management did not confide in Mr. Ford that the company was losing money at the tractor's retail price of $585. Over the years, Ford lost approximately $9 million while producing tractors for Ferguson to distribute, while Ferguson made about the same amount. Ferguson's investment was a few typewriters and a distribution headquarters in the middle of Ford's Rouge plant. Only after Ford Motor Company was reorganized in 1946 did the management approach Ferguson with a new tractor plan to share the profit based upon investment. Ferguson obviously did not accept this proposal and the Gentlemen's Agreement collapsed.

Vintage Ford Tractors covers in detail the many developments of Ford and Ferguson that I was involved in, and I recommend it to all those interested in Henry Ford's great desire to serve the farmer by producing a tractor that would do for the farm community what the Model T did for transportation.

Ford 8N assembly line

Ford and the Development of the Farm Tractor

*I think it is safe to eliminate the horse,
the mule, the bull team, and the woman,
so far as generally furnishing motive
power is concerned.*
—W. L. Velie, Deere & Company, 1918

Tractor revolutionaries
Above: *Harry Ferguson, left, and Henry Ford conspired to revolutionize the farm tractor. Photographed at the 1939 press introduction of their Ford-Ferguson 9N, the two men examine their novel tractor with Ferguson's radical three-point hitch and draft control system. The tractor world would never be the same again.* (From the collections of Henry Ford Museum & Greenfield Village)

1953 Ford NAA
Left: *Framed in the window of an abandoned barn, the classic lines of the Ford NAA tractor can be clearly seen.*

If you grew up on a farm, you are probably experienced with Ford tractors. Tractors by Ford have been milestones in the history of world agriculture. The venerable Fordson and its progeny forged a legacy of affordability, dependability, and capability that set the standard for farming.

This book is about the first fifty years of Ford tractors, about the Ford men and women who shaped the company, and about the times when these tractors were new.

The need for farming machinery grew out of the dreams of the pioneering farmers in the Great Plains and the far West of North America. Fledgling farm machinery firms spawned the first, huge, steam-powered machines in North America. These big tractors were capable of handling twelve or more plow bottoms, and were the prime movers for the monstrous early threshers. At that time in North America, 93 percent of the population lived on farms, but only a few on big farms. Most Americans and Canadians subsisted on farms of forty acres (16 hectares) or less while Europeans often lived on even smaller farms.

These small farmers in North America lived in the styles of houses immortalized by Grant Wood and Grandma Moses. Heat was provided by a cast-iron stove. Evening and morning light came from kerosene (paraffin) lamps. Fewer than 10 percent of the homes had indoor plumbing. Alexander Graham Bell invented the telephone in 1876, but few found their way into rural homes until after World War II. Henry Ford's Model T automobile became the farmer's friend in 1908, providing the farm family with a new freedom and mobility.

Life on these small farms depended on the horse. The farmer usually had one horse for every ten acres (4 hectares) under tillage. The horse was more than just a farm animal like a cow or pig, and more than just a pet like a cat or dog; a camaraderie developed between farmer and horse. The horse toiled along with the farmer to provide food for all. And they both shared in the rewards, with each horse annually eating the output of about three acres (1.2 hectares). Because horses needed to eat some of what they produced, the opportunity for the small tractor arose.

The first decades of the twentieth century saw tractors grow to gigantic proportions. Huge Cases, Rumelys, Hart-Parrs, and big International Harvesters ruled the fields in the United States and Canada; Saundersons and Marshalls worked Great Britain; Renault in France; Fiat in Italy; Lanz and Deutz in Germany; and Caldwell Vale and Jelbart in Australia all tilled the soil. Only farmers with great spreads of land could use or afford these monsters. And horses were still needed, as these behemoth tractors were only useful for plowing and belt work.

The birth of the gas-engine-powered farm tractor is generally credited to two individuals named Charles—Charles Hart and Charles Parr. They made the first successful production tractor in Charles City, Iowa, in 1902. In doing so, they started the internal-combustion traction engine industry and coined the term "tractor."

A gold rush was soon on to build ever smaller, lighter, and more useful tractors. Case had been in the steam traction engine business since 1885, but in 1913 added internal-combustion tractors. International Harvester introduced tractors in 1906. Massey-Harris and John Deere bought existing tractor companies to add to their list of farm implements in 1917 and 1918, respectively.

It was into this world that the Fordson arrived in 1917. It was small and cheap, designed for the small farm. Henry Ford had been raised on a small farm, and his interest in tractors, he said in an interview, was "To make farming what it ought to be, the most pleasant and profitable profession in the world." For the next fifty years, tractors from Ford were tailored to the needs of small farmers around the world.

These, then, were the times that gave birth to the classic Fordson, Ford-Ferguson, Ferguson, and Ford tractors in North America and Europe.

1944 Fordson Model N P4 diesel
After World War II started in 1939, the Fordson paint color was changed from orange to green to lessen the visibility of the Fordson as a target for enemy aircraft.

Chapter 1

Henry Ford and the Roots of the Ford Tractor

History is more or less bunk.
—Henry Ford, 1916

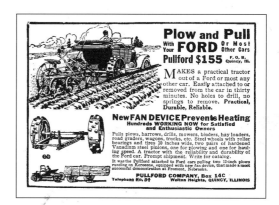

Above: **Pullford tractor-conversion ad**

1914 Ford Model T
Left: *The automobile that forever changed the shape of the world. Henry Ford launched his Model T in fall 1908. It was not the first car nor the best nor the least expensive, but it was a good, simple car, and it was mass produced. It soon became almost synonymous with the word "automobile."*

In 1879, a sixteen-year-old youth by the name of Henry Ford strode into Detroit in search of his fortune. Ford had walked some eight miles to the big city from the family farm in nearby Dearborn. At the time, Detroit had been continuously part of the United States for less than seventy years, and Michigan had been a state for only forty-two years. The American Civil War had ended just fourteen years earlier, and General George Custer had made his famous last stand just three years previously. When Henry Ford was growing up, an acre of rich Michigan land could be bought for about a day's pay. Timber wolves still lurked in the forests around farms. Deer and turkey were hunted for food, and maple trees provided sugar and syrup, as white sugar was not generally available.

Henry was born not into poverty, but into a comfortable seven-room home his father, William Ford, had built for his family. In 1832, William and his parents had immigrated from Ireland via steerage to escape the Irish potato famine. Michigan was good to William and he prospered.

In 1858, William was doing carpentry for a fellow Irishman, Patrick O'Hern. O'Hern had come to Canada with the British army, deserted, and escaped to Detroit. He built a home in Dearborn, married, and adopted a daughter, Mary Litigot. When William arrived to do carpentry, he met Mary O'Hern, and they were married in 1861.

Henry was born some thirty years after the family arrived in Michigan, on July 30, 1863. There was a stillborn child before Henry; three brothers and two sisters followed. The family, including the O'Herns, lived in the frame house that William built.

As young Henry grew, he assumed duties around the farm. There is no question that farm work of the time was hard; as Henry succinctly states in one of his surviving journals, "There's just too much work around the place."

Henry never took to farm labor. He had an especial dislike for chickens, saying "the chicken is fit only for hawks." Later in life he said that his goal in creating his farm tractors had been "to lift the burden of farming from flesh and bones and place it on steel and motors."

By the time Henry was seven years old, he and the other farm children of the area went to a one-room schoolhouse after the farm work settled down for the winter. *McGuffey's Eclectic Readers* bolstered the morals, conscience, and self reliance that parents had already instilled. The press of farm work resulted in sporadic schooling, however, leaving Henry a poor written communicator and a worse speller all of his life.

The hired men on the farm also influenced young Henry's fascination with things mechanical. A hired hand made Henry a gift of clock and watch gears, and Henry began not only to take things apart but also to put them back together. The Ford family soon had to guard watches, clocks, spring-wound toys, and the like from Henry's curiosity. The youth soon became proficient at clock and watch repair, and he claimed he fixed his first watch at age thirteen—and made the tools he used in the repair. Tools always interested Henry: "They were my toys," he wrote in his 1922 autobiography, "and they still are."

In 1876, when Henry was not quite thirteen, his comfortable childhood ended abruptly when his beloved mother died in childbirth. Henry had been close to his mother. He later said that the family home seemed "a watch without a mainspring." After becoming one of the world's richest

Westinghouse steamer

Henry Ford saw this steam-powered road locomotive when he was a twelve-year-old boy, and it helped set the direction of his whole life. Later, in 1913, when he had become a successful automobile magnate, Ford would go to great trouble to locate this exact engine, by then a rusty hulk. The owner wanted $10 for it, which Henry paid—but he also gave the owner a new Ford Model T as a bonus. Henry had the engine restored to like-new condition and again threshed with it on his sixtieth birthday. (Ford Motor Company)

men, he was asked what factors contributed to his success. He replied, "I have tried to live my life as my mother would have wished."

A few months after his mother's death, Henry was riding to Detroit in a horse-drawn carriage with his father. On the way, they met a steam road engine pulling a loaded wagon. Henry later recalled that this event was the greatest experience of his young life. "I was off the wagon and talking to the engineer before my father knew what I was up to," Ford wrote. "It was that engine which took me into automotive transportation."

He eventually left the farm for Detroit and soon got his first job at James Flower & Brothers Machine Shop. While working in this and other machine shops, Henry made ends meet by doing watch repair in the evenings. One of Henry's fellow apprentices was David Buick, who would also go on to make a name for himself in the automobile business.

Henry Ford, Steam Engineer

After three years in Detroit, Henry returned to Dearborn to work for a neighboring farmer named Gleason, who had purchased one of the pioneering steam engines built by G. Westinghouse & Co. of Schenectady, New York. The neighbor had hired a mechanic, but the mechanic was afraid of the engine, and quit. Gleason then asked Henry to get the steam engine running—a challenge Henry could not refuse. Ford said, "I was as proud as I have ever been when he asked if I might run the engine."

Henry got the engine running, and then spent the summer as its engineer. He threshed grain and clover and sawed lumber with it for eighty-three days straight. "I became immensely fond of that machine," Ford later wrote, "and I have never been better satis-

Henry Ford and Quadracycle

Thirty-three-year-old Henry Ford and his first automobile, the Quadracycle. The pioneering automobile had a two-cylinder engine, no reverse gear, and no brakes. On the front was a doorbell to warn away errant pedestrians. Ford later used his Quadracycle to attract investors to his first automobile company. (Ford Motor Company)

fied with myself." And after his summer working with the Westinghouse engine, Henry signed on as a Westinghouse traveling repairman.

In December 1884, Henry took a major step toward becoming an automobile magnate: he enrolled in Goldsmith's Bryant & Stratton Business University. Here he studied mechanical drawing, accounting, and business. This was to be his only formal business education. It must have been sufficient, for it was said that Henry Ford was never bested in a business deal.

One reason for Henry's attempts at improving himself with a college education may have been a bright-eyed, chestnut-haired, twenty-year-old neighbor girl that he met at a dance. Her name was Clara Jane Bryant, and the pair would be married in 1888.

During winter, Henry went to business school. In summer, he was on the road for Westinghouse. Besides servicing Westinghouse engines, Henry often fixed other steamers as well, thus acquiring extensive background in steam engine design. One halcyon day in 1885, he was called to repair an Otto gasoline engine in Detroit. This was Henry's first exposure to the novel four-cycle internal-combustion gas engine that German inventor Nicolaus Otto had patented in 1876. "No one in town knew anything about them," Ford wrote, "there was a rumor that I did, and although I had never been in contact with one before, I undertook and carried out the job."

By 1886, Henry was tiring of life on the road. To keep Henry on the farm, William Ford offered his son an eighty-acre (32-hectare) farm he owned, half of which was not cleared. The farm included a small house that would suit a young couple. The house, linked with the prospect of setting up a steam-powered sawmill on the property, appealed to the twenty-three-year-old Henry.

While logging his land, Henry worked with a black man, William Perry. While they shared a crosscut saw, Henry gained respect for the hard-working Perry. Later, when the Model T was in full production, Perry became a foreman, and blacks were given equal status to whites in the Ford factories, an uncommon situation in auto companies at the time.

When Henry was not working his sawmill, he was experimenting with gas engines. In 1891, Henry decided to return to Detroit where he was offered a job at the Edison Illuminating Company. Henry saw the move as an important step for his future: he desired to understand electricity, which he saw as the key to making gas engines work properly. Also, the timber on his farm was gone.

Henry started at Edison as an engineer-mechanic at forty-five dollars per month—not a bad wage in 1891. When the generators at his sub-station were up and running, Henry had the time and place to continue his experiments. The chief electrical engineer would call Henry before tearing any unit down for repair; Henry would come to help, even on his own time.

By then, Henry's obsession with the gas engine was to integrate it into a self-propelled farm vehicle. In a little machine shop he erected in his power sub-station,

AutoPull tractor-conversion ad, circa 1910s

Numerous firms—both legitimate and questionable—offered kits and plans for converting Ford Model T cars into tractors of sorts. More than forty conversion kits were on the market from 1908 to 1917. Some worked quite well, others did not. (Smithsonian Institution)

Henry began work on a small, single-cylinder gas engine. On Christmas Eve 1893, Henry brought the engine home. Anxious to continue his experiments, he clamped it to the kitchen sink. Clara was pressed into service, even though she was expecting her parents for dinner. While Clara dribbled gas into the intake, Henry spun the flywheel. Suddenly the little engine burst into life, belching flames, noise, and smoke, and nearly shaking the sink apart. That Clara put up with these noisy, smelly, and dangerous antics on Christmas Eve with guests due says a lot about her supportive nature—especially when their six-week-old son Edsel was sleeping in the next room.

Around this time Henry was hired by the Detroit YMCA to teach evening machinist classes. One of his students was a brilliant young man named Oliver Barthel. Barthel and his employer, Charles B. King, were wildly enthusiastic about the prospects of a gas-motor-driven carriage and were experimenting with a prototype. In Henry, they found a kindred spirit. When King's motor wagon took to the street in 1895, Henry provided escort on a bicycle. Later, when Ford assembled his first car, the Quadracycle, King gave him four valves for the engine.

Dawn of the Horseless Age

At the turn of the century, the newly founded *Horseless Age* magazine reported on all known motor wagon experiments—and starting in 1895, there were quite a few. The French had been first, in 1894, but the Duryea brothers had the first American auto running, in 1895.

Building a motor carriage, or car, became a fad among the mechanically inclined of the period. From 1900–1910, more than five hundred companies were formed to manufacture cars. One inventor that *Horseless Age* failed to recognize was George Seldon of New York, a patent attorney who patented the horseless carriage in 1895. This meant that all other inventors were infringing on his patent and were liable for royalties. It is doubtful that Seldon was a serious auto maker, but he saw the coming industry and wanted a piece of it. Seldon eventually sued Ford, but Ford won and nothing more was heard of Seldon.

In 1896, the year after Seldon's patent went into effect, Ford and Edison friends Edward "Spider" Huff and James Bishop completed Ford's first car. Called the Quadracycle, it was crafted in the woodshed of Ford's home.

The 500-lb (225-kg) auto was completed in the wee hours of the morning of June 4, 1896. Ford and Bishop were ready for the roll-out but discovered the car would not fit through the door. Ford was not deterred; he took an ax to the door frame and bricks of the rented structure. Once the car was outside, the engine was cranked up, and with a wave from Clara, Ford was off, with Bishop giving chase on a bicycle.

The Ford Motor Company later overcame Seldon's patent claims by maintaining that Ford worked on the Quadracycle several years earlier than he actually did. It was only when modern historians dug into the details that the truth came out.

Henry Ford, Automotive Engineer

The Irish of Detroit formed a somewhat informal society. Detroit Mayor William Mayberry, for example, was an acquaintance of Henry and his father. Another influential Irish acquaintance was William Murphy, who had wide interests in Detroit, including a substantial share of Edison. Murphy was also a horseless carriage enthusiast. When he saw Ford's Quadracycle, he challenged Ford to make a sixty-mile (96-km) trip around Detroit; if the Quadracycle could make the trip without accident or breakdown, Murphy would take him seriously.

For the next three weeks, Ford prepared his auto, then showed up at Murphy's residence. "Let's go for a ride," said Ford. The trip was completed without incident, and Murphy offered financial backing for an auto firm to be called the Detroit Automobile Company. Besides Murphy and others, Mayor Mayberry invested in the new company. Established in 1899, the firm was the first car manufacturer in Detroit, and Ford was listed as Mechanical Superintendent.

Almost from the start, things were not well at the company. Instead of building the peppy and reliable Quadracycle, the first product was a delivery van, and the company dissolved. The reason for the failure was that Ford withheld his genius; he was not in charge, nor was his share of the payoff what he thought it should be.

The losses may have discouraged Mayor Mayberry but apparently did not deter Murphy from working with Ford. Soon, five of the Detroit Automobile Company investors led by Murphy incorporated the Henry Ford Company. Ford was given one-sixth of the stock and the title of Chief Engineer. The investors had learned little about working with Ford.

It was not long before Ford was up to his old tricks. But Murphy called in an acquaintance Henry M. Leland as a troubleshooter. Just four months after formation of the new company, Ford was sacked. Murphy lost little time in rechristening the firm as the Cadillac Automobile Company. Ford's two-cylinder engine was replaced by a single-cylinder unit built by Leland, and the Cadillac gained an unparalleled reputation for reliability and quality. It later became the flagship of the General Motors Corporation.

Leland and Ford would cross automotive paths again, in 1921. Leland founded the Lincoln Automobile Company after GM took over Cadillac. When Lincoln foundered in the hard times of the early 1920s, Ford bailed it out. He then summarily fired Leland.

In late 1902, Ford formed a partnership known as Ford & Malcomson Ltd. with a wealthy Detroit coal dealer, Alexander Malcomson. Malcomson was an operator in every sense of the word. He had brought his coal business up by its bootstraps on borrowed money, but despite his success, Malcomson did not have much cash, certainly not enough to put Ford into the car business. He did have credit, however. To keep his creditors from knowing that he was taking a flyer into automobiles, Malcomson put his clerk, James Couzens, in charge of the venture, opening accounts in Couzens' name. The new Ford car created by this part-

Growing of food, making of tools, and transportation are the three basic jobs!
—Henry Ford

nership was the Model A, which was to be robust and practical.

Auto companies in those days were mainly assembly and marketing operations. Sub-assemblies, such as engines, transmissions, and bodies, were purchased from outside vendors. Ford & Malcomson arranged with several suppliers for parts for their Model A. Sales did not materialize, however, so the bankers and suppliers decided their best hope of collecting was to take stock in the car company.

Thus, on June 16, 1903, papers incorporating the Ford Motor Company were signed with little fanfare. Stockholders included Couzens and engine and transmission suppliers John and Horace Dodge among others. The stockholder list was rounded out by Henry and Malcomsen, who were assigned stock for machinery, patents, and designs. Between them, the duo held 51 percent of the 1,000 votes.

Even with this backing, the new company was literally down to its last nickel when Chicago dentist Dr. E. Pfennig bought the first Model A. In the next eight months the Ford Motor Company sold 658 automobiles, and the stockholders shared a dividend of $98,851.

It wasn't long before Ford and Malcomson locked horns over whether to build luxury or low-priced cars. Along with the Model A, the firm offered the Model B and Malcomson's luxurious Model K. By the first stockholders' meeting, Ford saw himself and Couzens on one side and "those fellows" on the other. When the smoke cleared, Ford and Couzens had won; the others resigned and turned in their stock. By this gambit in 1905, Ford wound up with 585 of the 1,000 shares of stock.

Henry Ford, Automotive Magnate

The most significant automotive event of the twentieth century occurred in 1908 with the introduction of the Ford Model T car. When the Model T debuted in the fall, it was a sensation. There were cheaper cars, but none so advanced as the Ford. It had a good engine, a planetary semi-automatic transmission, and a battery-less ignition system. The frame was made of vanadium steel, which allowed the car to be stronger but lighter than previously possible. By the end of World War I, half the motor cars on earth were Model T Fords.

In 1913, Ford began experimenting with a moving assembly line. At first the moving line was just for building flywheel magnetos, but by January 1914, the whole car was assembled as it moved down the line.

Also in January 1914, Ford announced that production workers' pay would be boosted from $2.34 to $5.00 per day. Furthermore, Ford created three work shifts; to staff the extra shift, some 5,000 new employees were needed. The pay raise was so dramatic as to be almost unbelievable; the newspapers called it an impossible act of generosity. Within days, there were 15,000 desperate men milling about in front of the Highland Park Ford plant seeking jobs.

The highly touted $5.00 daily rate was not nearly as simple as it sounded. Daily pay actually stayed at $2.34; the $5.00 figure included a profit-sharing bonus that had to be earned by following certain rules: workers had to have at least six months of service; they had to be at least twenty-two years old unless married or supporting a widowed mother;

some of the bonus had to be saved or invested; and they had to live the temperate, wholesome lifestyle Ford mandated. Women were not included, at least at first. Ford always did pay women better than other firms did, however. "We expect the young ladies to get married," said Ford: "I pay them well so they can dress attractively and get married."

To enforce the bonus conditions, inspectors from the newly instituted Ford Sociological Department made the rounds of employees' homes. Each inspector had a new Model T, chauffeur, and interpreter appropriate to the neighborhood he was visiting. Wives, family members, and neighbors were questioned: Ford bonus money would not be squandered on wild living or extravagance.

Nevertheless, the $5.00 day paid off in several ways. Among other benefits, Ford employees could now afford to buy Ford cars. And as other companies raised their pay to remain competitive, their employees also could afford automobiles.

The brothers Dodge, stockholders in Ford, were not happy about the $5.00 pay rate. They thought Ford was giving profits away just to spite them since they were double-dipping as suppliers and stockholders. Their pique probably gave Ford the idea of how to get rid of stockholders altogether.

In 1914, Ford unveiled the much-publicized repayment of $50 to every Model T purchaser that year—the first rebate. What Ford was really after, by removing $11 million from profits, was denying the Dodge brothers $1 million they had been counting on. The Dodge brothers were peeved.

Furthermore, Henry was talking about investing in 2,000 acres (800 hectares) of marshy land beside the River Rouge for a super plant that would include a harbor where Ford's own navy would bring in iron and coal. Blast furnaces would make steel. Cargo ships would ferry finished cars and tractors across the oceans of the world. Sub-assembly suppliers like the Dodges would no longer be needed.

The Dodge brothers launched a lawsuit. Meanwhile, Henry bought the Rouge site with his own money and in his own name; he also incorporated a new company on July 27, 1917, Henry Ford & Son, to manufacture tractors at the new plant. A judge ordered Ford to pay the Dodges and other stockholders a fair share of dividends. Henry responded with appeals and delaying tactics. The stockholders offered to sell Henry their stock for a fair price, but he said he had enough stock.

Things went along normally until December 30, 1918, when fifty-five-year-old Henry Ford announced that he was resigning the Ford Motor Company presidency. Son Edsel took over. Henry and Clara took leave of Detroit for California. Shortly thereafter, the *Los Angeles Examiner* ran an article stating that Henry was starting a new car company to compete with Ford. No one wanted stock in the old Ford firm. After a few weeks, there were some mysterious inquiries and low-ball offers for stock. The trail led back to Henry. Eventually, he was able to buy all outstanding shares for $106 million, and all talk of a new car company was dropped. Ford Motor Company was once again firmly in Ford's hands. Henry Ford was back in business.

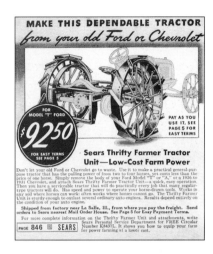

Sears Roebuck tractor kit
Even the great mail-order house Sears Roebuck offered a tractor conversion kit so you did not let your Ford car "go to waste."

For most purposes, a man with a machine is better than a man without a machine.
—Henry Ford,
My Life and Work, 1926

Chapter 2

Dawn of the Fordson Tractor

The new-fangled tractors will be the ruination of the farmer because they don't make no manure.
—Tractor non-believer's proverb, 1920s

Above: **1937 Fordson Model N**

1926 Fordson Model F
Left: *As a young farmer, Fred Bissen used this 1926 Fordson F and six horses to farm his 240 acres (96 hectares) near Adams, Minnesota. He has since restored the tractor.*

Despite his distaste for farm chores, Henry Ford was always fascinated by agriculture. In his later life when he was successful and wealthy, he would state, "Growing of food, making of tools, and transportation are the three basic jobs!" Ford's genius in creating his farm tractors was to combine these "three basic jobs" in his Fordson and later Ford tractors. All along, Ford had faith in his vision of a gasoline traction engine, or tractor, as he said in 1926: "I felt perfectly certain that a tractor could do the hard work of pulling a plow, and that horses . . . did not earn their keep."

In 1906, Henry Ford began to experiment with gasoline tractors. By 1907, he demonstrated his Automobile Plow machine. This light tractor used the engine and transmission of the Ford Model B car. The four-cylinder engine of 284 ci (4,652 cc) developed 20 hp. Many of the other components came from the heavyweight six-cylinder Ford Model K that ex-stockholder Malcomson had insisted on building. At least Ford was finding good use for them. But Ford's Automobile Plow proved to have too little traction as well as overheating problems when pulling a load. Twelve more variations of the Automobile Plow were built in 1908 and experiments continued.

Ford used his new wealth to buy up farm land around Dearborn. On the plots, he continued to test tractor prototypes built with vanadium steel and Model T components. Ford also loved to show how farmers could use the Model T to drive their buzz saws or corn huskers with a belt run from the car's jacked-up rear wheel.

As Ford's tractor plans became public knowledge, other entrepreneurs stepped onto the stage. A group from Minneapolis lead by financier W. Baer Ewing realized that the name "Ford" was magic, and rushed to incorporate its own Ford Tractor Company, in 1916. This company really did include a man named Ford—one Paul B. Ford—but he and the tractors the fledgling firm created bore no relation to Henry Ford nor his tractors. It is not known whether the "Minneapolis Ford" was conceived as a way to sell tractors based on name confusion or in the hopes that Henry Ford would pay royalties to use his own name on his own tractors. Minneapolis Ford tractors did not sell; these poorly engineered tractors subsequently bore the dubious honor of being the impetus for the University of Nebraska Tractor Tests when a disgruntled owner, Nebraska state legislator Wilmot F. Crozier, instigated the tests to guard against false claims made by tractor makers. In addition, Henry Ford did not buy his name back from the group, choosing instead to organize his tractor company in 1917 as Henry Ford & Son.

Between 1906 and 1916, Ford spent about $600,000 of his own money on tractor experiments. By summer 1916, prototypes of a new Ford tractor were shown to the public. Joe Galamb, the gifted Hungarian chief engineer who had helped Henry with the automobile planetary transmission and the moving assembly line, was sacked: the tractors he designed looked too much like the Model T. The new tractors were no longer Model T derivatives, but were the entirely new concept that Ford was looking for. The new chief engineer was Eugene Farkas, also of Hun-

1906 Ford Automobile Plow
Henry Ford's early, 1906 version of the Automobile Plow used the four-cylinder engine and transmission from the luxurious Ford Model B car. (From the collections of Henry Ford Museum & Greenfield Village)

garian decent.

Farkas chose the unit frame concept pioneered by the innovative Wallis Cub tractor. The engine block, transmission housing, and rear-axle housing served double duty as the frame of the machine. The new, larger engine was provided by the Hercules Engine Company of Evansville, Indiana; it displaced 251 ci (4,111 cc) and fathered some 20 hp. A selective-shift, three-speed transmission was used. The first derivation, prior to the public showing, had left-hand steering like a car and no engine hood. For the public, the steering shaft was moved to run down the right side of the engine; this placed the driver in a better position for guiding the wheels in the plow furrow. The engine was also covered with a louvered hood. These would be the historic first tractors to bear the name badge "Henry Ford & Son."

Ford's design team had the benefit of the convenient test fields of Ford's newly acquired Fair Lane estate, named for the Cork, Ireland, home of father-in-law Patrick O'Hern and designed by the young architect Frank Lloyd Wright. The designers also had the advantage of being able to test their tractor against just about every other tractor make. Ford, however, was still not ready to commit to production. He wanted to wait until the tractor was "as right as Lizzie," referring to the Model T's pet name.

The M.O.M. Tractor Saves the Day

Prior to the introduction to the U.S. press, two "Henry Ford & Son" tractors were sent to England for testing by the British government, which was embroiled in World War I. Because Britain, and much of Western Europe, had come to rely on low-cost imported grain from the United States, Canada, Australia, and Russia, British farmers had moved to livestock production.

All of [the competitor's tractors] impressed us as too heavy and very much underpowered, so that became our first problem to solve. We also found that most of these vehicles had some outside form of drive open to the dirt that was thrown up in farming operations. So, we decided that our tractor's drive would be inside a housing just like that on a motorcar.
—Charles Sorensen on creating the Fordson, *My Forty Years with Ford*, 1956

Henry Ford riding in the 1907 Ford Automobile Plow

Above: *Henry's later, 1907 version differed from the previous by having an automobile-type radiator. The axles and differential came from a Ford Model K six-cylinder car as seen in the background.* (From the collections of Henry Ford Museum & Greenfield Village)

1907 Ford Automobile Plow

Right: *Henry's 1907 Automobile Plow never went into production. This example has been preserved at the Henry Ford Museum & Greenfield Village.*

With the outbreak of World War I, German blockades and U-boats threatened grain shipments to Great Britain. Along with mobilizing its troops to fight the Germans, the British government was forced to mobilize farmers for food production. At the same time as it commandeered farm horses for the British Army, the government established tilling goals for each arable acre. To achieve these goals, the British Board of Agriculture ordered British tractor makers to work overtime, as records indicate there were just five hundred tractors in Great Britain in 1914. The board also called for the importation of all the American machines available. But there still were not enough tractors to fulfill the tillage demands.

Fortunately for Ford and the British, Lord Percival Perry was a member of the Board of Agriculture. In 1906, Perry had been involved in selling three Ford cars to be used as the first London taxis. He and his wife had gone to Detroit to negotiate for what would become British Ford Motor Company. Ford invited the couple to move in with the Ford household during their stay, and while there, the Perrys and Fords became quite fond of each other. Perry later became a Lord, the head of British Ford, and a Board of Agriculture member, and in May 1916, during Britain's time of need, he arranged for the testing of Ford's prototype tractors.

A government committee was amazed by the Ford tractor's small size, light weight, durability, ease in starting, and good handling. The committee recommended that British Ford start immediate production. Since Ford already had been granted a license to set up a Model T factory in Cork, Ireland, it was decided to produce the tractors there also. A very pleased Henry Ford donated the drawings and patent rights for the tractor for British war production.

1914 Ford prototype tractor
Engineer Joseph Galamb created this 1914 experimental Ford tractor based heavily on the Model T car. One of the cylindrical tanks held fuel, the other held additional cooling water for the radiator. The Model T planetary transmission was used; this could be the first tractor to have a Torque Amplifier–type down shift. (From the collections of Henry Ford Museum & Greenfield Village)

However, the German bombing of the British homeland changed plans. Tractors were needed immediately, and if the factory in Ireland was bombed, further delays would be unavoidable. Thus, the British Ministry of Munitions (M.O.M.) placed an order for 6,000 Ford tractors from the United States. Ford cabled back that "they would work day and night, and comply with every request."

Ford launched a crash development program. Prototype tractors were built in pairs, with each succeeding pair incorporating improvements over the last. The trac-

tors were continuously subjected to exhaustive testing on the Fair Lane grounds. Records are sketchy, but probably eight pairs of what have become known among collectors as X Models were made, a total of sixteen

M.O.M. tractor prototypes.

The British were concerned. When would the design be frozen and deliveries start? The British Lloyd George government had an expediter in Washington named Lord Northcliff, and he made a trip to Dearborn. With a campaign of flattery, Northcliff got Ford off top dead center and into production. Finally, 254 of what have become known as M.O.M. tractors were delivered in 1917.

While these M.O.M. tractors shared the same specifications as the Fordsons, they did not bear the Fordson name. The name "Fordson" came as a result of the M.O.M. program because much of the coordination was carried out in Morse code via the transatlantic cable. To speed up communication, abbreviations and acronyms were used whenever possible. Thus, "Henry Ford & Son" became "Fordson." Henry Ford liked the name and chose it for production tractors when the M.O.M. program was over.

In addition to the Ford-built tractors, the Ministry of Munitions imported tractors from the Emerson-Brantingham Implement Company of Rockford, Illinois, and J. I. Case Threshing Machine Company of Racine, Wisconsin, as well as the General Motors' Samson tractor, the Waterloo Boy made by Deere & Company of Moline, Illinois, and the Titan made by International Harvester Company of Chicago, Illinois. The tractors were distributed throughout Great Britain by the County War Agricultural Committees, which also hired drivers; bought and delivered fuel, oil, and parts; and kept records of costs and production. The County Kent Committee reduced the number of models it operated to three: the Waterloo Boy, Titan, and Ford. By the end of the 1918 plowing season, the Ford wore the crown for the lowest cost per acre of plowing for fuel and oil, the best acres-per-hour plowing rate, and less downtime for repairs. In the end, the program was a great success for Great Britain as wartime food production met demand.

The M.O.M. order got Ford's tractor program off to a good start, although it was a hectic time at the firm's Dearborn plant. Records show that Henry Ford & Son initiated its system of mass producing tractors, and at the same time, made a profit on the program.

When the armistice was signed on November 11, 1918, ending the "war to end all wars," Ford was in a fine position to enter the North American tractor field.

Launching the Fordson Tractor

Just as he had amazed the world with his revolutionary Model T car, Henry Ford again stunned everyone with the introduction of his radical Fordson farm tractor. The Fordson was conceived as a lightweight, inexpensive tractor to replace not the gigantic steam engines of the large farms but the mules, oxen, and horses of the more numerous small farms. The timing was right, the $785 price was right, and the Fordson became the Model T of farm tractors.

During World War I, the U.S. Department of Agriculture controlled tractor production, and it granted a license on June 26, 1918, to Henry

As early as 1912, Henry Ford conceived the idea of a tractor and plow as a single unit. Like many simple things, it apparently had not been thought of before. The horse and plow had been separate, and so, in the reasoning of the day, the tractor that supplanted the horse should be hitched to the plow. The unit tractor-plow was part of our plan when we organized to build the Fordson in 1915.
—Charles Sorensen, *My Forty Years with Ford*, 1956

1910s Ford Farkas prototype

Above: *One of the first versions of engineer Eugene Farkas's tractor design. This prototype represented a complete departure from the earlier Model T–based tractor designs. The sole Model T part was the coil box behind the engine. A Hercules four-cylinder engine was used.* (From the collections of Henry Ford Museum & Greenfield Village)

1910s Ford experimental motor cultivator

Left: *This prototype tractor incorporated Model T, as well as Fordson, parts.* (From the collections of Henry Ford Museum & Greenfield Village)

1916 Ford Farkas prototype

Above: *The steering wheel is now on the right for better plowing visibility. Note the same type of seat is used as on the Galamb experimental tractor. Note also that the worm drive is above the differential. Heat from this inefficient unit made driving the tractor uncomfortable.* (From the collections of Henry Ford Museum & Greenfield Village)

1917 Ford Farkas prototype

Right: *This was the twenty-eighth example of the Farkas prototype tractor, fitted here with an Oliver plow. The oval badge on the radiator reads "Henry Ford & Son."* (From the collections of Henry Ford Museum & Greenfield Village)

1917 M.O.M. X Model

Facing page: *The Ministry of Munitions (M.O.M.) prototype X-9. Pre-Fordson X Models were built in pairs, and there were probably sixteen tractors in all. Each pair progressed to become more like the eventual production Fordson. The worm gear set was below the differential on these machines, eliminating the problem of the hot seat on the Farkas.*

Above, top: **1919 Fordson ad**

Ford executives and 1919 Fordson

Above, bottom: *Executives from Henry Ford & Son stand proudly with an early Fordson in March 1919. Edsel Ford is on the far left, Henry Ford is in the center with the fedora hat, and Charles Sorensen is to the right of Henry. (From the collections of Henry Ford Museum & Greenfield Village)*

Ford & Son to manufacture tractors for the American market. Typically, a quota of 1,000 tractors for each agricultural state was established. During the war years, farmers wanting to purchase new tractors had to obtain a permit from their local County War Board. Fordsons were sold through regular Ford car and truck dealers, although the Henry Ford & Son firm was still separate from the Ford Motor Company.

The British Ministry of Munitions order for tractors had started production in fall 1917 and continued into 1918. The M.O.M. tractors can be discerned from the later Fordsons by several identifying parts. Fordsons had a Fordson logo on the front of the gas tank as well as on the radiator and toolbox whereas M.O.M. tractors had no logos. Fordsons had the words "Manufactured by the Henry Ford & Son, Dearborn, Michigan" on the back (driver-end) of the gas tank whereas M.O.M. tractors had no writing there. The gas tank on M.O.M. tractors was also rounder than Fordson tanks and had no top center seam. Otherwise, the M.O.M. tractors were much like the pre-1920 Fordsons, and the last half of the M.O.M. order for 6,000 tractors was filled during Fordson production, in 1918.

The first ten Fordsons were delivered to special friends of Henry Ford. The first official Fordson Model F, as the version following the M.O.M. order was called, was built for botanist Luther Burbank, a close friend of Henry Ford. It rolled off the line on April 23, 1918. The Hercules engine company had supplied some unnumbered engines for replacements of M.O.M. engines that did not pass muster for some reason. The first ten Fordsons used these replacement engines, which were hand stamped with serial numbers 1 through 10.

After these first ten tractors, the Fordson numbering system began with serial number 6901. This leaves a gap from the end of the M.O.M. numbering, which had begun with serial number 1 through number 3900. Ford probably jumped to a significant base serial number for record-keeping purposes.

Between serial numbers 6901 and about 7260, produced in the last weeks of April 1918, Fordson F models and M.O.M. tractors were mixed on the assembly line, although they each kept their own characteristics. Tractors in this sequence can be either type of tractor, depending upon whether they were for the U.S. and Canadian markets or for the Ministry of Munitions order.

When the Fordson F was launched in 1918, the implement industry at first did not take it seriously. The Fordson was small and light. At 2,700 lbs (1,215 kgs), it was dwarfed by other tractors of the times, such as Deere's 6,000-lb (2,700-kg) Waterloo Boy or the 8,700-lb (3,915-kg) IH Titan. Also, Ford was not in the farm implement business and sold the tractor through car dealerships. Most tractor suppliers also made equipment such as threshers, spreaders, and mowers. The implement industry failed to anticipate Ford's powerful distribution network and

the low price Ford could offer because of its manufacturing capacity.

A total of 34,167 Fordsons were built in 1918. With the end of World War I on November 11, 1918, plans to build the Fordson in Ireland were revived. The first Cork-built Fordson rolled off the line on July 4, 1919.

Henry had further plans up his sleeve. At the end of 1919, he established Eastern Holding Company, part of Ford's gambit for acquiring all the minority stock in the Ford Motor Company. With family capital of more than $100 million, Eastern Holding procured the assets of the Ford Motor Company, Henry Ford & Son, and some of Ford's other business interests. In early 1920, when these activities were completed, the Eastern Holding name was dropped in favor of the Michigan-chartered Ford Motor Company.

Without stockholders holding the reins, Henry was now able to work toward his goal of manufacturing tractors, cars, and trucks from raw materials, rather than by purchasing sub-assemblies. One of his first moves was to build the giant River Rouge plant outside Detroit.

The Rouge plant was a marvel. It had its own powerhouse, steel smelting and rolling plant, casting and stamping plant, and ship-docking facility. Henry then bought his own coal and iron mines, and Ford-owned ships carried these commodities to the Rouge. Fordson production ended at the Dearborn plant and was moved to the new Rouge plant. Thus, for Fordsons made in the United States from the 1920 model and on, the stamping on the end of the fuel tank read "Manufactured by Ford Mo-

1917 M.O.M. tractor and Ford executives
Ford executives posed on October 6, 1917, with the first M.O.M. tractor to be shipped to England. Ernest Kanzler and Eugene Farkas are in the front row at far right. Henry Ford is in the center with Charles Sorensen on his left. (From the collections of Henry Ford Museum & Greenfield Village)

1920 Fordson
Above: *An early publicity photograph showing an early, 1920 production Fordson tractor. Note the solid-sided radiator shell. Until 1923, Fordsons didn't have brakes. In 1923, a brake integral with the clutch was added; the brake was activated by the clutch pedal.* (From the collections of Henry Ford Museum & Greenfield Village)

Reins for your Fordson
Left: *The Fordson's competition came not from other tractors but from horses. For those farmers who were unsure of the newfangled steering wheel, Rowe Manufacturing Company offered its Line Drive that allowed farmers to drive the Fordson with reins.*

tor Company, Detroit, Michigan."

For the 1920 model year, Ford began supplying the Fordson's engine. It was essentially the same as that which had been supplied by Hercules, a 251-ci (4,111-cc) four-cylinder unit rated at 1,000 rpm. The tractors were painted machinery gray from the beginning, but bright red wheels were added in 1920. Fenders were not yet offered, although aftermarket fenders were available. Also, the ladder-side radiator was replaced with a solid-side unit in 1920.

Hard Times in the Tractor Business

A severe economic downturn surprised everyone in 1920. In the previous year, Ford had sold more vehicles than ever before as there had been considerable pent-up demand for cars, trucks, and tractors following the war, which gave Detroit several boom years. Worried about inflation, the U.S. government then cut spending by $6 billion, which caused the boom to end suddenly. The Ford Motor Company was particularly unprepared. Ford had just spent $60 million on the River Rouge plant, some $20 million on the Dodge brothers settlement, and another $15 million to buy coal and iron mines. These expenditures, combined with the buy out of the stockholders, left Ford in serious threat of bankruptcy.

The crisis sharpened Henry's wariness of bureaucracy in the Ford Motor Company. He had chief auditor Louis Turrell make a complete survey of the company, identifying cost-saving possibilities. When the task was done, Turrell fired his staff. The next day, Ford fired Turrell. Then extra desks, cabinets, telephones, typewriters, and the like were sold.

Ford's penny-pinching methods had little effect. Ernest Kanzler, Edsel's brother-in-law and manager of the tractor operation, did have an idea that raised substantial cash quickly. During the war, Kanzler streamlined the inventory system to eliminate stockpiling that tied up space and dollars. Railroad cars brought in materials needed for the day and took away completed tractors. When the crisis of 1920 hit, Kanzler was called upon to do the same for the Ford car plant. Kanzler found some $88 million worth of parts on hand. To cut inventory, he had packages of spare parts made up that were sent to dealers along with their car shipments—and these car shipments also included some unordered cars. Any dealer refusing to accept and pay cash for a shipment risked loosing their franchise. At that time, a Ford franchise was like a license to print money and replacement dealers were easy to get. Few dealers hesitated to pay, going to their own banks to borrow the money. In that way, the dealers financed Ford through the worst part of the crisis.

For the Fordson tractor operation, the crisis worsened in 1921. There were then 166 companies in North America building around 200,000 tractors per year. Deere & Company had entered the tractor market in 1918 by purchasing the Waterloo Gasoline Engine Company, maker of the Waterloo Boy tractor. Deere had scheduled to construct forty tractors per day for the last half of 1921. In the end, however, Deere sold only seventy-nine tractors in the whole year. With the economic down-

The Fordson on the farm arose
Before the dawn at four:
It milked the cows and washed the
    clothes
And finished every chore.

Then forth it went into the field
Just at the break of day,
It reaped and threshed the golden
    yield
And hauled it all away.

It plowed the field that afternoon,
And when the job was through
It hummed a pleasant little tune
And churned the butter, too.

For while the farmer, peaceful-
    eyed,
Read by the tungsten's glow,
His patient Fordson stood outside
And ran the dynamo.
—Anonymous poem, 1920s

1925 Fordson Model F
By 1925, when this Fordson was new, 75 percent of the world's tractors were Fordsons. This tractor is fitted with a two-bottom plow developed specifically for the Fordson by the Oliver Chilled Plow Company. Owner: Don Artman of Monee, Illinois.

turn, farmers were among the first to feel the effects. Rather than taking the risk of a newfangled tractor, most just decided to keep old Dobbin at work.

If this was a problem for Deere, Ford had scheduled production of 300 Fordsons daily for 1921. When production was up to speed at the River Rouge plant in 1922, production was discontinued at the Cork, Ireland, plant and at the newly instituted Hamilton, Ohio, plant. By the time the seriousness of the recession was realized, the yards around the Rouge plant were filled with Fordsons.

Tractor Price Wars
Henry knew what to do. He first cut the Fordson's $785 price to $620. Other tractor makers followed suit, so Ford further cut his price to $395. This drastic price cut marked the end of the line for a large number of the independent tractor makers, who soon closed shop. Even the giant General Motors withdrew its tractor entry, the Samson, which it obtained when it bought the Samson Sieve Grip Tractor Company of Stockton, California. Ford, however, managed to sell 35,000 Fordsons in 1921, 67,000 in 1922, and more than 100,000 in 1923.

In the wake of the tractor price war, the competition included only about ten manufacturers of any consequence. Deere's reaction to the Fordson challenge was to compare its Waterloo Boy to other tractors in the field—and they found it wanting. International Harvester had introduced two new standard-tread tractors, a 15/30 in 1921 and a 10/20 in 1922. Hart-Parr Company of Charles City, Iowa, had bowed a new 12/25 and 15/30 in 1918. All of these new tractors followed the style set by the Fordson rather than the antiquated look of the Waterloo Boy. All were smaller, lighter, had automotive-type steering, front-mounted radiator and fan, and an engine hood like that of a car. Before Deere's acquisition of Waterloo Gasoline Engine Company of Waterloo, Iowa, Waterloo engineers had been working on a modern version of the Waterloo Boy. Deere engineers quickly dusted off the blueprints and developed the tractor into the first production two-cylinder tractor to be called a John Deere, the Model D introduced in 1923. To say that the Model D was a success is an understatement: it was ultimately in production for thirty-one years.

The Fordson price cuts also awakened the sleeping giant, International Harvester. IH rushed to develop a machine to best the Fordson. Ford's tractor did not have a driveshaft power takeoff (PTO), and therefore was not suitable for the new harvester implements. While it could replace some horses on a farm, it could not replace them all because the Fordson was not useful for cultivating crops. By 1924, IH was ready with its contender, the all-purpose Farmall. Sales of the Farmall exceeded expectations in 1925, and by 1926, IH's new Rock Island, Illinois, plant was in operation, and Farmalls were rolling out the door.

The Farmall marked a revolution in agriculture. It differed radically from all other tractors of the time in that its small front wheels were set close together to run between two rows. Its rear axle did not run straight between the rear wheel hubs, but was connected to the hubs via a large gear mesh. The result was a tractor with a high rear axle, providing 30 inches (75 cm) of ground clearance. The rear wheels were larger in diameter than normal and were wide enough apart to straddle the two rows that the front wheels ran between. The Farmall was powered by a 20-hp engine and weighed about 4,000 lb (1,800 kg). Its main custom implement was a two-row cultivator mounted to the front frame, which for the first time in a volume-production tractor offered power cultivation of crops such as corn and cotton.

The tractor price war of the 1920s allowed many small farmers to get into power farming. Weaker companies were eliminated from the tractor market when they could not lower their prices to compete. The survivors were forced to copy Ford's production-line methods and rede-

THE FORDSON AGRICULTURAL TRACTOR

THE FORDSON—whose reputation is established—is now available through a local dealer . . . available embodying the following distinctively modern tractor features.

An engine that develops 30 horse-power at one thousand revolutions per minute. High-tension magneto with enclosed starter coupling to insure easy starting. Hot-spot manifold and carburetor for gasoline. Cooling system with water pump driven by a V-type fan belt. An air washer that holds 17 quarts of water.

Other features include a filter to separate grit and carbon from the oil in the lubrication system. Transmission fitted with large, roller bearings. Large gear has double bearing mounting.

Sixteen-plate, multiple-disc transmission brake with increased plate surface. Rear-wheel bearing lubricated automatically. Gears that shift easily from increased release movement.

Coil-type front spring. Heavier rear-wheel fenders and platform. Heavy, sheet-metal steering wheel. Heavier, stronger, one-piece front wheels. Still other features optional at extra cost are: a fly-ball governor, pulley, lighting system powered by a generator driven from the fan belt, extra long cleats for rear wheels, and extension rims.

All these features together with reliability, economy and long life. A local dealer will demonstrate its farm advantages to you.

FORD MOTOR COMPANY

1922 Fordson ad

We had farmed with mules until one day Dad came home with a new Fordson. We had two ground-driven binders of about the same size. He hooked one to the Fordson and the other to a team of four mules. He got on the Fordson and gave me the binder and the mules and we headed for the field. During that day, I and the mules passed the Fordson four times!
—Farmer Charles Domeier of Pecatonica, Illinois

sign their products to be cheaper and more appealing to smaller farmers. In the long run, the Fordson's competition had a healthy effect on the industry.

Both the Fordson and the Farmall marked major advancements in the development of the farm tractor. Just as the Fordson changed the concept of the conventional, heavyweight tractor to that of the standard-tread, lightweight machine, the Farmall further refined it to that of the all-purpose row-crop configuration. At International Harvester the combination of standard tractors and the new Farmall soon gave IH a 70 percent share of the tractor market. After 1926, the North American market clearly called for the Farmall style of general-purpose tractors.

Ford transferred all Fordson production to Cork since the European market was not yet affected by the demand for GP tractors. Henry said he needed the factory space in the United States for the new Model A car anyway. Fordson production in Detroit ended in mid-1928. Nearly 850,000 Fordsons had been delivered in ten years of production.

Fordson Model F Developments

The appearance of the Fordson F changed little during its years of production. A lighter shade of gray paint was used starting in 1924. Fenders were made an option that same year. In 1925, the toolbox fenders, sometimes called orchard fenders, were made available. Besides being a handy place to store tools, these fenders were supposed to prevent rearing accidents for which Fordsons were becoming infamous.

When the transmission brake was added in 1923, the use of industrial Fordsons with hard-rubber tires abounded. Even GM owned one with a front-mounted sweeper.

Between 1924 and 1926, more than 25,000 Fordsons were shipped to Russia. During 1925, the 500,000th Fordson was built, and an unsurpassed tractor-model annual production record of 104,168 was set.

Many aftermarket variations on the Fordson theme appeared during the mid-1920s, including graders, road rollers, and golf course mowers. One of the most unique was the Trackson, built by the Full Crawler Company of Milwaukee, Wisconsin. This conversion made the Fordson into a track-layer, which was popular with loggers. More than 88,000 Fordsons were converted to Tracksons before Full Crawler was bought in 1930 by Caterpillar Tractor Company of Peoria, Illinois.

No new models of cars, trucks, or tractors were introduced by Ford in 1927. GM, and especially its Chevrolet division, was coming on strong as a competitor, and Ford car and truck sales were declining rapidly.

Henry Ford's presence in the implement province and the new type of competition he soon introduced returned the industry for a time to the atmosphere of battle.
—Cyrus McCormick III, *The Century of the Reaper*, 1922

1925 Fordson ad

Therefore in 1927, Ford focused its technical talent on creating the new Model A car line. In mid-year, Ford car and truck production was terminated, leaving most dealers with little to sell—except for Fordson tractors. Thus, in its final North American production year, which actually ended in 1928, 101,973 Fordsons were built.

The Irish Fordson Model N

In December 1928, a new Ford company was created in Great Britain. Ford transferred control of the British Ford Motor Company, Ltd., of Dagenham, England, and Henry Ford & Son, Ltd., of Cork, Ireland, to British citizens and to a new firm, Ford Motor Company, Ltd. The new company was assigned worldwide rights to the Fordson tractor, and in 1929, all production tooling was transferred to Cork.

Because of the worldwide depression in the 1930s, the European car and truck markets were hurting. Lord Percival Perry, head of the new British Ford, needed cash flow and even the spare parts business for existing Fordsons would be a worthwhile endeavor. Further, the Soviet Union had tried to order more Fordsons to augment the 25,000 they already had, but was turned down when Detroit production was terminated. The Soviets then turned to International Harvester and placed a substantial order. Perry reasoned that more Soviet orders might be forthcoming, and he would be ready if they did.

Early in 1929, production of the Fordson began in Cork. The versa-

1924 Fordson Model F

By 1924, the price of the Fordson had risen from its low of $395 in 1922 to $495. This price was held through 1925. Fenders and belt pulley were optional at extra cost. Owner: Charles Hardesty of Valparaiso, Indiana.

1926 Fordson Model F
Above: *The boat-tail toolbox fenders made their debut at the end of 1926. These fenders replaced the square-tail fenders of the early models. Owner: Fred Bissen.*

1926 Fordson Model F
Right: *The Fordson F was rated at 10 drawbar hp and 20 hp on the belt. The difference in horsepower was due to belt slippage and to the power required just to propel the tractor itself.*

tility, traction, and reliability of the new all-purpose tractors from IH, Hart-Parr, Deere, and others had doomed the Fordson in the North American market. It needed a complete update, similar to the changes from the Model T to the Model A car. Ford had considered upgrading the Fordson, or even creating an all-new tractor, But before any definite decision was made, however, the world's economy was hit with the stock market crash of 1929 and the Great Depression.

To make the Fordson acceptable in the European market, Perry and his team decided to upgrade the Fordson within its basic design. The reworked Irish Fordson was named simply the Model N. The engine displacement was increased to 267 ci (4,373 cc) over the 251 ci (4,111 cc) of the Fordson F by widening the bore 0.125 inches (3.125 mm). Rated operating speed was increased from 1,000 rpm to 1,100 rpm. The F's Model T magneto and coils were replaced by a high-tension impulse-coupling conventional magneto. The Model N was available in either a 21-hp kerosene or 26-hp gasoline version.

The spoke front wheels of the F were replaced by cast wheels. The Model N also had a heavier front axle with a downward bend in the middle and a larger water-wash air cleaner. The paint scheme remained the same lighter gray color initiated in 1924, and the wing-type orchard fenders were standard. On the rear end of the fuel tank were the words "Ford Motor Company, Ltd., England, Made in the Irish Free State," although some read "Made by Ford Motor Company, Ltd., Cork, Ireland." Finally, the Model N was heavier than the Model F by almost a half ton, weighing in at 3,600 lb (1,620 kg).

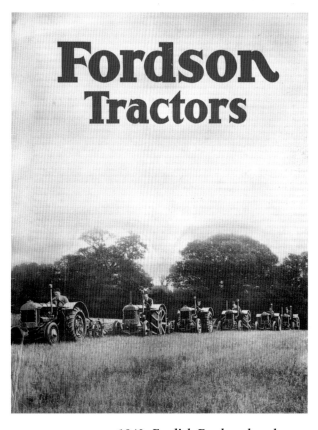

1940s English Fordson brochure

The English Fordson Model N and All-Around

In 1929, British Ford began building a new factory on the outskirts of London in Dagenham, Essex. Because of the continuing depression, there was excess production capacity when the factory was completed in 1932. At the time, Cork Fordson production hit a low of fifty tractors per week.

Only 31,471 tractors had been built in the four Irish model years of 1929–1932. Henry had originally wanted to build tractors in Cork due to his need to give something back to his roots. But now there were new political problems caused by the separation from Great Britain of southern Ireland into the Irish Free State. Few of the Fordsons built in Cork stayed in Ireland anyway, and the cost of raw materials was high and availability of skilled labor low. The cost of the Cork operation, coupled with the excess capacity in Dagenham, caused Perry to disregard Henry's sentimentality. Thus, from August 1932 through February 1933, the Cork production facilities were dismantled and shipped to Dagenham.

During the six months the facilities were being relocated, the Fordson

1922 Fordson Model F Hadfield-Penfield crawler

This Fordson was converted to a half-track, a popular arrangement at the time. Hadfield-Penfield Steel Company of Bucyrus, Ohio, made these crawler conversion tracks that virtually doubled the weight and drawbar pull of the basic Fordson. Owner: Gene Runkle of Ostrander, Minnesota.

was given another styling facelift. The tractors were now painted a striking blue, conventional fenders with the toolbox on the dash were added, a ribbed pattern was cast into the radiator tank, and the Fordson name was cast in the radiator side panels. The fuel tank end now read "Ford Motor Company, Ltd., Made in England." The Model N designation was retained. Only 2,807 Fordsons were produced in 1933 because of the Great Depression and the move to Dagenham.

The Model N received another updating in 1935. The upright steering wheel was replaced by a new wheel that was angled at 40 degrees. A new double-leaf seat spring replaced the single-leaf type, and a rear-output PTO was available. Orange trim was added to the blue paint. Pneumatic tires were an option, as were lights and a starter.

The first variation on the basic Fordson configuration debuted in 1937. The Fordson celebrated it's twentieth birthday with a new model, the tricycle-configured Fordson All-Around. The All-Around was an attempt to get in on the move toward general-purpose row-crop tractors. Production was then split between All-Around and regular Model N versions. Exporting All-Arounds to the United States and Canada was also an effort to regain some of the North American market that was lost when American production ceased.

Oil-bath air cleaners were first used on Fordsons in 1937, replacing the old water wash air cleaner. Higher-compression heads for both the

gas and distillate models arrived for the 1937 models as distillate, or Tractor Vaporizing Oil (TVO), was just coming on the market. It allowed the use of higher compression than did kerosene, and therefore, more power was produced without increasing displacement or operating speed. The paint color was also switched to all orange.

The onset of World War II forced a changed to green paint on the 1939 model, as orange tractors were considered easy targets for German Luftwaffe pilots. Green remained the Fordson's color, except for the many military variations, until production of the Model N ended in 1945.

Eugene Farkas' low-cost tractor design had lasted twenty-nine years, from the first M.O.M. tractors of 1917 through the final Fordson Model N and All-Around of 1945. It was a production record exceeded only by the John Deere Model D, produced for thirty-one years from 1923 through 1953. It was quite an accomplishment for a tractor that was dismissed by the experts of 1917 as being "too light to stand up." The Fordson was loved by some, hated by others. It nevertheless had such a profound impact on world agriculture that, today, almost any farmer over fifty years old has an opinion about them. The Fordson certainly did fulfill Henry Ford's ambition to "lift the burden of farming from flesh and bones."

1931 Fordson Imperial row-crop
An Imperial conversion of a Fordson for row-crop cultivation. The conversion was done in 1931 by the Farm Tractor & Equipment Company of Des Moines, Iowa. The truck is a 1931 Ford. (From the collections of Henry Ford Museum & Greenfield Village)

1937 Fordson Model N engine

Above: *Among the improvements over the Model F that were incorporated in the N was the regular-impulse magneto, replacing the flywheel-type magneto and coil box. The N used an upgraded engine with the bore increased from 4.00 to 4.125 inches (100 to 103 mm). This gave a displacement of 267 ci (4,373 cc) rather than the 251 ci (4,111 cc) used previously. In addition, a 100-rpm increase in operating speed gave the N about 20 percent more power.*

1937 Fordson Model N

Right: *This Model N was built in Dagenham, Essex, Great Britain, during the production run from 1933 to 1946. The blue-and-orange paint scheme was adopted in 1935. Cast front wheels and a downward-bent front axle characterize the N. Radiator shutters, or a curtain as shown here, were necessary for proper operation with kerosene (paraffin). Owner: Eric Coates of Southampton, Hampshire, Great Britain.*

1939 Fordson Model N

Above: *Model Ns were given a bright all-orange paint job starting at the end of the 1937 production run. Also incorporated at that time were higher-compression heads and an oil-bath air cleaner. Rubber tires became an option on Fordsons in 1935. Owner: Dennis Crossman of Walton, Somerset, Great Britain.*

1938 Fordson Model N Howard Dungledozer

Left: *The Howard Rotavator Company of Great Britain offered its Howard Dungledozer attachment for Fordsons.*

1944 Fordson Model N P4 diesel

Facing page: *The engine design of the Fordson dated all the way back to circa 1917, and the engine was always a weak point for the later models. Installation of a Perkins diesel gave a large power increase, longer life, and lower fuel consumption. This N has the popular postwar Perkins P4 conversion. Owner: Jim Spark of Romsey, Hampshire, Great Britain.*

1938 Fordson All-Around

Above: *The steering mechanism was revised on the 1938 Fordson All-Around, with the "chicken roost" bar lowered and reinforced so it was more rugged. Orange paint was used for parts of the 1938 and 1939 model years before Fordson switched to the wartime green. All-Around production continued through 1940. Owner: Marlow Remme of Dennison, Minnesota.*

1937 Fordson All-Around

Facing page: *By the twentieth birthday of the Fordson, row-crop tractors were big sellers in the North American market. The All-Around could accommodate mounted cultivators whereas the regular Fordson could not. The All-Around also incorporated steering brakes for sharper turns.*

Chapter 3

Harry Ferguson, Farm Tractor Genius

*Your Fordson's all right as far as it goes, but it
doesn't really solve the problem.*
—Harry Ferguson, 1917

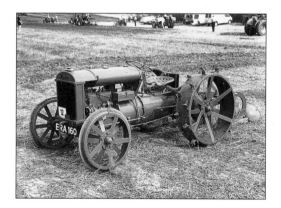

Above: **1935 Ferguson-Brown Type A**

1923 Fordson Model F with Ferguson plow
Left: *Harry Ferguson crafted the ingenuous Ferguson
floating-skid-mounted two-bottom plow for the
Fordson tractor. Note the two-point hitch arrangement and the depth linkage.*

It has always puzzled agricultural historians why humans, despite wit and creativity, have been so slow to improve farming tools. It could be because God cursed the ground when Adam sinned (Genesis 3:17), or it could be that farmers were so busy growing food that they had no time to work on improvements. Whatever the reason, it was 1952 before tractors outnumbered horses on American farms.

Horses, oxen, and mules pulled farming tools from earliest times. When tractors arrived in the nineteenth century, they were merely mechanical alternates for the tractive power of animals. There was no basic change, just a substitution. There technology rested until Harry Ferguson appeared on the scene.

Harry Ferguson was a complex man. In his biography of Ferguson, *Tractor Pioneer*, Colin Fraser said Ferguson "combined extremes of subtlety, naiveté, charm, rudeness, brashness, modesty, largess and pettiness; and the switch from any one to another could be abrupt and unpredictable."

Although he could be unpredictable, Ferguson, however, was unwavering in his struggle to perfect his concept of integrating the farm tractor and its implements into one unit. This was the famous Ferguson

1930s Ferguson-Brown Type A
While onlookers watched attentively, Harry Ferguson in fedora and high boots demonstrated the plowing ability of his Type A.

System on which he worked, beginning in 1917, for twenty years before it began to pay off. And pay off it did: The 1939 Ford tractor with the Ferguson System made Harry Ferguson a multimillionaire. Since then, more than 85 percent of the tractors made worldwide have incorporated his system, although most of their trade names make no reference to him.

Ferguson was born in Ireland on November 4, 1884. Ireland, the Emerald Isle, incorporates two countries, a division formally made in 1920. The northeastern fifth of the island is called Northern Ireland and is part of the United Kingdom of Great Britain and Northern Ireland. The independent Republic of Ireland makes up the rest of the island.

Northern Ireland has a harsh climate and rugged terrain. When Ferguson was born in Growell, Northern Ireland, the country was called Ulster, and the inhabitants Ulstermen. Most Ulstermen were descendants of the Scots, transplanted to Ireland by King James I in the late 1600s; they were a Protestant minority in a Roman Catholic country. The intransigent character of the Ulsterman grew out of this background and was further tempered by adversity of climate and society. The result was "Ulster grit," of which the Ulsterman was rightfully proud.

Ferguson was born Henry George Ferguson, but was always called Harry. He was the fourth of eleven children of James and Mary Ferguson. James, the embodiment of Ulster grit, farmed 100 acres (40 hectares) in County Down. With the relatively large farm, the Fergusons were better off than most, but because of the harsh conditions, the large family survived only by unremitting toil.

Harry Ferguson had a propensity for confrontation with both his father and schoolmasters. He left school at fourteen to work for his father on the farm. He grew to hate horses and the hard work. His small stature and light build made handling the plow especially arduous for him.

Harry Ferguson, Engineer

Ferguson's oldest brother, Joe, left farming to become an apprentice mechanic for a linen spinner in 1905. After a few years, his intense interest in the budding motorcar field inspired him to open a repair garage for autos and motorcycles. Harry was also fascinated by these mechanical contrivances and soon was working for Joe.

During the years he worked for his brother's successful repair business, Harry attended the Belfast Technical College. Besides his technical ability, he developed an uncanny talent as a promoter and salesman. Early on, he recognized the importance of image in promoting a business, which led him into motorcycle and car racing to publicize his brother's garage.

By 1908, Harry had another interest: aircraft. He and a friend, John Williams, made trips to several air meetings, where they made measurements of existing flying machines. Back in Belfast, Harry convinced brother Joe that building and flying the first airplane in Ireland would be good for business.

Ferguson's monoplane was finished in mid-1909. Engine and propeller troubles plagued progress, preventing anything but short bounces into the air. Finally, on December 31, 1909, Ferguson was determined to fly in order to be on record as having flown that year. Despite horrible weather conditions, fly he did—for about 130 yards (119 meters), after which he made a successful landing.

The budding British aviation world was not ready to recognize as a colleague one from the bogs of Ireland. Ferguson suffered a barrage of criticism in the fledgling *Flight* magazine. Therefore, when a £100 prize was offered by the Irish town of Newcastle to anyone making a three-mile (4.8-km) flight there, Ferguson entered. As it turned out, his was the only entry. Again, unfavorable weather conditions prevented success. Ferguson made many attempts, each one ending in a crash and damage to the craft, but not to Ferguson himself. The press was having a field day with more criticism. Demonstrating true Ulster grit, Ferguson persevered. He went through three propellers, two wings, three wheels, and sundry other parts. The press began referring to him as the "plucky Irish Aviator." Finally, Ferguson moved his takeoff base to a more favorable spot. He got airborne, flew past Newcastle, and on to the prize and credibility.

A later crash demolished the craft and severely injured Ferguson. The airplane was repaired, only to be crashed and destroyed by his friend Williams. The seat and the eight-cylinder J.A.P. engine were all that were preserved and are now in a Dublin museum.

Racing and flying had made Ferguson quite famous in Ireland. It

In 1917, while in England, I met Ferguson, then a young machinery salesman. I told him that we proposed to combine the [tractor with the plow into one unit]. He took up the idea and in a few weeks came back with some models. . . . Had I been able to foresee the consequence of that meeting I would have avoided it.
—Charles Sorensen, *My Forty Years with Ford*, 1956

also led to resentment from brother Joe and an eventual parting of the ways. Harry then started an automobile business of his own, May Street Motors. One of his first hires was a twenty-year-old natural mechanic, engineer, and designer named Willie Sands. Sands would be Harry Ferguson's right-hand man into the 1950s, and his contributions to Ferguson's success were fundamental.

Harry Ferguson Revolutionizes the Farm Tractor

In 1914, Ferguson established a corporation known as Harry Ferguson, Ltd., and obtained a franchise for the Overtime tractor, which was the renamed British version of the American Waterloo Boy. When World War I loomed, Ferguson was appointed by the Irish Board of Agriculture to oversee tractor maintenance and records for all of Ireland as part of the crash program to increase food production that spawned the M.O.M. tractor program.

Ferguson and Sands traveled throughout Ireland in a government-supplied car, helping farmers set up plows, and aiding them in getting the most out of their tractors. All the while, they collected records and determined which features were the most desirable. Ferguson later reported that plows were the main problem, not tractors. To get three or four plow bottoms adjusted to give even furrows was a major accomplishment. Up to that time, most plowing was done with a single, hand-controlled horse plow.

The tractor supplied by Henry Ford & Son had its own problem: rearing. After only several months of M.O.M. tractor plowing, the first driver was killed when his tractor reared. The plow struck a rock, the deep-cleated wheels could not slip, and the tractor spun over around the rear axle, pinning the driver to the ground. The Fordson later became so noted for this tendency that aftermarket clutch depressors and ignition interrupters for it were sold. Finally, in 1924, Ford offered the tailed, or orchard, fenders designed to pop the cleats out of the ground and allow slippage before the tractor went over. Ferguson, however, saw the problem in the plow design, not with the tractor.

Ferguson set about creating a lightweight mounted plow for a tractor. By 1917, the Model T Ford was famous all over the world, and enterprising industrialists were capitalizing on its popularity with farmers by offering as many as forty-five different kits for converting the car into a tractor. One of these was the Eros conversion from the E. G. Staude Manufacturing Company of St. Paul, Minnesota; it was one of the best, and became quite popular in England and Ireland as well as in North America. It seemed obvious to Ferguson that the common tractor plows of the day were much too heavy for the Eros. With the help of Sands, he designed a mounted plow that Harry Ferguson, Ltd. would manufacture. When completed, the plow weighed only 220 lb (99 kg), about half the going rate for a two-bottom plow, and it had only half the number of parts.

More importantly, the plow featured an ingenious attachment to the Eros that hitched it under the tractor's belly, forward of the rear axle. Thus the force of plowing was reflected downward on all four wheels,

increasing traction. It also eliminated any tendency for the tractor to rear up if the plow caught on a solid object. The plow was mounted with weight-compensating balance springs that allowed the driver to raise and lower it with a lever beside the seat. Even though the Eros tractor enjoyed only a brief production life, the Ferguson plow sold quite well.

Just as Ferguson's Eros plow hit the market in 1917, Ford announced it was setting up its Cork tractor factory. Also announced was the 6,000-unit M.O.M. order for imported tractors. Ferguson might have greeted the Ford-built tractor with resentment as it spelled the end of his Eros plow. But not one to miss opportunity, Ferguson saw the M.O.M. tractors provided an even larger plow market.

In fall 1917, when Ford emissary Charles Sorensen came to London to discuss setting up production in Cork, Ferguson and Sands rushed to meet him, plow drawings in hand. "Your Fordson's all right as far as it goes," Ferguson told Sorensen, "but it doesn't really solve the fundamental problem."

Sorensen was somewhat put off by the remark, but Ferguson had his attention. Ferguson then rolled out his Eros plow drawings, and using his gift of persuasion, convinced him of the desirability of his underbelly line of draft. Ferguson's earnestness, enthusiasm, and confidence won over Sorensen. The meeting ended with Sorensen's commitment to support development of the Ferguson plow for the new Ford tractor.

Ferguson's Duplex Hitch Plow and Draft Control

Ferguson and his design team, which now included Sands, John Williams, and Archie Greer, began designing their Fordson plow. To overcome the tractor's rearing and to transfer draft loads to all four wheels, a duplex linkage consisting of two parallel links mounted one above the other was employed. This arrangement was semi-rigid in the vertical plane, but allowed movement in the lateral plane for steering.

Ferguson and Sands traveled to Dearborn in 1920 to show the plow to Ford and Sorensen. Ferguson intended for Ford to manufacture plows by the millions, paying Harry Ferguson, Ltd., royalties. Ford misinterpreted Ferguson's motives and had Sorensen offer him a job. When Ferguson refused several escalating salary offers, Ford was piqued. Ford then tried to buy patent rights to the plow. Again Ferguson refused, but with a smile. Both Irishmen had met their match in stubbornness. At that, they parted, having established nothing more than a grudging, mutual respect.

Back in Belfast, Ferguson and his team, now minus Sands who was trying his hand at his own business, tried to overcome the tendency of the duplex hitch plow to make a furrow of uneven depth. Trailer-type plows of the time used a depth wheel, mounted behind the plow and running in the furrow, to keep depth constant. Several trips were made to the Cork Ford plant in 1921 where tests were conducted with the tractor. One of the people detailed to help was a shell-shocked ex-prisoner of war named Patrick Hennessey, for whom tractor work was considered therapeutic. Hennessey would later become British Ford Motor Company chairman, receive knighthood, and have many dealings with Ferguson.

Ferguson then got the Roderick Lean Company of Mansfield, Ohio, to manufacture the Fordson plow in the United States, but there was still depth-control problems with the plow. The depth wheel prevented the draft loads from bearing down on the tractor's rear wheels, resulting in insufficient traction. Ferguson rehired Sands to work on the problem.

By 1923, Sands had the solution. It was called the floating skid, and a patent was obtained. A skid plate, which replaced the depth wheel, was connected to the duplex hitch via linkages. If the tractor's front wheels went over a ridge, tending to force the plow to run deeper, the linkages forced the skid plate downward, raising the plow to its original depth.

In 1924, the Roderick Lean Company went bankrupt, so Ferguson made a manufacturing arrangement with George and Eber Sherman, who were Fordson distributors for the state of New York and confidants of Henry Ford. The firm of Ferguson-Sherman, Inc., was formed with a plant in Evansville, Indiana, to make the plow.

The duplex hitch plow was a success and sales were brisk. Like the Eros plow, the Fordson duplex hitch plow was strictly mechanical. The floating skid worked well enough for the plow, but now Ferguson and his team were working on other implements, such as harrows and cultivators.

Hydraulics were next on Ferguson's agenda. His team began to work with hydraulics and added a third link to the duplex hitch. The draft load reacts against a heavy spring in this third, or upper, link, actuating a hydraulic valve. The implement automatically raised or lowered from its preset position as draft load increased or decreased. In this way, although implement depth varied slightly, the draft load was kept constant. If a plow running eight inches (20 cm) deep, for example, encountered hard soil, the increasing draft loads caused the plow to raise to about six inches (15 cm) until the hard going was passed. Then the plow was automatically returned to its original depth. In addition, the act of raising the plow dramatically increased the download on the rear wheels, reducing slippage.

This concept, which was variably called "weight transfer" or "draft control," was patented in the United States. The full patent title was "Apparatus for Coupling Agricultural Implements to Tractors and Automatically Regulating the Depth of Work," and it was granted in 1926.

But then the plucky Irishman was dealt another blow: Fordson production in North America ended in 1926.

The Black Tractor and the Ferguson-Brown

Faced with the onset of the Great Depression in 1929, Ferguson realized there was little chance of new tractors being developed, and less likelihood of getting his system incorporated into a current tractor. But Ferguson was not to be outdone by the times. Instead, he designed a tractor from scratch with the hydraulics built in, not added on. With drawings in hand, Ferguson contacted companies specializing in the required parts. Parts were purchased and a tractor assembled.

The first prototype Ferguson tractor was a diminutive machine that bore more than a passing resemblance to a Fordson. It weighed less than 2,000 lb (900 kg), just a little more than half that of the Fordson. As with the Fordson, the tractor was powered by a Hercules engine whereas the transmission and differential were supplied by the British gear manufacturer David Brown. Ferguson's tractor was painted black, and as it had no other name, it came to be known as the "Black Tractor."

The Black Tractor had its share of shortcomings, but it performed up to par. Ferguson persuaded David Brown to manufacture an improved version known as the Ferguson-Brown Type A, which was ready for sale in May 1936.

Sales of the Ferguson-Brown were disappointing, and the tractor was never a success, according to Ferguson's chief field engineer Harold Willey (who would later be hired by Ford). The tractor's price was too high, especially since new implements were also needed. Brown thought the tractor was just too small. Brown and Ferguson had a falling out, and Brown began on his own to make the changes he thought necessary. Ferguson's influence continued in David Brown's tractors through 1972, when the firm was bought by Case.

With that, Ferguson crated up a Ferguson-Brown Type A with implements and sailed for the United States. He had asked his friends the Sherman brothers to get him another audience with Henry Ford.

On the ship, Ferguson encountered Leon R. Clausen, the crusty president of J. I. Case Threshing Machine Company. Not one to pass up an opportunity, Ferguson offered Clausen first chance at his weight-transferring system. Clausen spurned the offer, saying real tractors didn't need it. They were words he would live to regret.

My father had an old Austin tractor and six horses on the farm. But when he got his Ferguson he was able to get rid of four horses and sell the old Austin as well.
—English farmer John Chambers, whose family owned the first 1936 Ferguson-Brown tractor

1936 Ferguson-Brown Type A
As Ford N Series designer Harold Brock remembered, "The Ferguson-Brown tractor, including Ferguson's implement hitch, proved to be an unsatisfactory design and was given but casual observation by Ford engineers."

Chapter 4

The Tractor of the Century

Our competition was not other tractor brands. We felt our competition, at least back in 1938, was the horse.
—Harold Brock, Ford 9N designer, 1990s

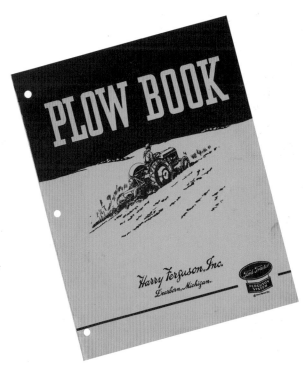

Above: **1940s Ferguson Plow Book guide**

1941 Ford-Ferguson 9N
Left: *The tractor of the century. This 9N is owned by Jack Crane of Whitestown, Indiana.*

What the world needs right now is a good tractor that will sell for around $250.
—Henry Ford, 1937

On July 30, 1938, Henry Ford celebrated his seventy-fifth birthday. The whole town of Dearborn turned out to celebrate with him. On the condition that alcoholic beverages never be tolerated there, Ford donated twenty acres (8 hectares) on the River Rouge, downstream from Fair Lane, for a municipal park christened Ford Field. He had suffered a slight stroke earlier in the year, but seemed to have recovered completely. However, some people believed his judgment suffered some from then on.

After shifting Fordson production to Ireland and the British Ford Motor Company in 1927, Ford again begin tractor experiments at Fair Lane. Howard Simpson was now chief engineer, and his tractor designs made use of Ford truck parts, especially the 1937 truck model with its 85-hp V-8 engine. Simpson's tractor had a row-crop front end and used car wheels, rubber tires, and even hubcaps. This model was shown to agricultural journalists in January 1938. An updated version was built in 1938 using an L-head six-cylinder engine with a wide front end arrangement. Also in 1938, a Ford engineer named Karl Schultz designed a strange three-wheel tractor with one-wheel drive that also used the V-8 engine. Ford engineer Harold Brock designed the experimental 9R chassis that included a worm-gear final drive and other features that would eventually become the basis of the 9N chassis. Experiments continued.

In fall 1938, Ferguson arrived at Fair Lane in the company of the Sherman brothers. A crated Ferguson-Brown tractor and implements had been shipped to New York, and a Sherman employee brought the crates by truck to Fair Lane. At the time, Ford was inspecting his mines in the Michigan Upper Peninsula, so Ferguson spent several days uncrating and preparing the machinery. The tractor was primed and readied down to the last detail. If nothing else, Ferguson knew how to put on a show.

The Handshake Agreement

It was a bright blue October day when Ford returned to Fair Lane and the demonstration began. Ford's own test driver, Ed Malvitz, took the wheel. He first plowed a couple of rounds, then switched to a harrow. This was followed by a three-row cultivator, and finally by a ditcher. All assembled agreed that the Ferguson-Brown handily outperformed both a Fordson and a Model B made by Allis-Chalmers Company of Milwaukee, Wisconsin. When plowing up a grade, both the Fordson and the Allis spun their wheels, while the weight-transferring Ferguson System automatically pressed down on the Ferguson-Brown's rear wheels.

Ford was ready to get down to business. He called for a table and chairs to be brought out of the estate's gatehouse kitchen. Ford and Ferguson seated themselves, and Ferguson demonstrated the system's function by using a spring-wound model.

Ferguson stressed both men's Irish heritage, both were raised on

1938 Ford-Ferguson prototype
This 1938 experimental tractor designed by Howard Simpson used a Ford flathead six-cylinder engine. No data exists on this machine, although it is known that six-cylinder engines were not offered in Ford cars and trucks until 1941. (From the collections of Henry Ford Museum & Greenfield Village)

1938 Ford-Ferguson prototype
This experimental tractor was assembled for testing in the winter of 1938–1939. The design shows many of the characteristics of the forthcoming 9N. (From the collections of Henry Ford Museum & Greenfield Village)

farms, and neither man enjoyed the drudgery of the toil. They shared an interest in bettering the small farmer's lot. Ferguson convinced Ford that a tractor they made together could replace the horse at a reasonable price and would bring undreamed-of benefits to mankind. The two men had an extraordinary idealism and hit it off in fine fashion.

First, Ford made the same faux pas he made at their first meeting: He offered to buy Ferguson's patents. Ferguson retorted, "Mr. Ford, you haven't got enough money to buy my patents, because they are not for sale to you or anyone else at any price."

"Well, you need me as much as I need you. So what do you suggest?" asked Henry Ford.

"A gentleman's agreement," responded Ferguson. "You stake your reputation and resources on this idea, I stake a lifetime of design and invention. No written agreement could be worthy of what this represents. If you trust me, I'll trust you."

A few more details were discussed. Ferguson would be responsible for the design. Ford would manufacture the tractor Ferguson designed. Ferguson would design and manufacture a line of implements and set up a dealer network to handle tractors and implements. The British Ford Dagenham plant would supply tractors under the same arrangement.

Tractor visionaries
Above: *Harry Ferguson, left, with seventy-six-year-old Henry Ford at the press unveiling of the Ford-Ferguson 9N tractor in June 1939.* (From the collections of Henry Ford Museum & Greenfield Village)

Henry Ford and 1939 Ford-Ferguson 9N
Left: *Having finished a plowing demonstration at the 9N's press introduction in 1939, Henry Ford doffs his farmer's straw hat. Both he and Ferguson took turns plowing at the press debut. Then eight-year-old David McLaren, standing in front, mounted the tractor and plowed furrows just as straight and even.* (From the collections of Henry Ford Museum & Greenfield Village)

Horses and Tractors on U.S. Farms

Year	Horses	Tractors
1910	23,934,000	1,000
1915	26,493,000	25,000
1920	25,199,000	246,000
1925	22,082,000	549,000
1930	18,865,000	920,000
1935	16,676,000	1,048,000
1940	13,932,000	1,567,000
1945	11,629,000	2,354,000

—U.S. Census Bureau

1939 Ford-Ferguson 9N

Aluminum hoods and grilles were used on the earliest 700–800 9N tractors, in 1939. This was done because the steel-stamping equipment was not ready in time. Charles Sorensen's son supposedly had a foundry capable of making the patterns and castings in time, but in alloy. This tractor is serial number 527, probably built during the first or second week of production. Notice that the grease fittings for the kingpins are in front, an arrangement used only in 1939–1940; for all others N Series, these fittings were placed on the back, out of harm's way. The smooth rear-axle hubs were also used only on 1939–1940 models, whereas the 8x32-inch (20x80-cm) rear tires were standard until 1942. Owner: Charlene Meyer of Gurnee, Illinois.

With that, the two men stood up from the table and shook hands, sealing their gentleman's agreement. Ford would build tractors with the Ferguson System. As protection in this handshake deal, it was agreed that either party could terminate the agreement at will and without explanation. Ford was truly impressed by Ferguson, and in the end, he not only agreed to the deal—which held greater profit potential for Ferguson—he also agreed to loan Ferguson $50,000 to start his operation.

Ford management, on the other hand, was apprehensive about the agreement with Ferguson. Ferguson knew Henry Ford's reputation for not taking stock in written agreements, and it was obvious to Ford executives that Ferguson had plotted to take Henry Ford by surprise in forging a business partnership based solely on a conversation and a handshake, according to Ford engineer Harold Brock. Details of the agreement were never fully disclosed to Ford management; Henry Ford simply told chief executive Charles Sorensen that Ford would design and develop the tractor and Ferguson would market it, Brock remembers. No other details were forthcoming.

With nothing in writing, Ford managers were concerned about the scope of the gentleman's agreement and how it impacted Ford Motor Company's working with Ferguson. Ferguson ultimately had great power and was pampered by management: If Ferguson disagreed with the managers, he simply went directly to Henry Ford and got what he wanted, according to Brock.

Secret Development of the Model 9N

In January 1939, Ferguson, his wife Maureen, daughter Betty, Sands, technical assistant John Chambers, and several others arrived in Dearborn. Ferguson and his family set themselves up in the Dearborn Inn, the world's first airport hotel, which Ford built along with the airport in 1924 when he was developing his famous Ford Tri-Motor airplane.

Before Ferguson and his entourage arrived, Ford had constructed three prototype tractors incorporating the Ferguson System. These were sent to Georgia for winter testing.

The Ford and Ferguson teams began design work in earnest. Ford's chief engineer was Lawrence Sheldrick; the team leader was Harold Brock. About ten engineers were installed in a remote corner of the Rouge plant called the Blue Room, where they would not be bothered. Sheldrick, Ford, Ferguson, Sorensen, and Chambers were personally involved on a

The tractor had been a mechanical substitute for Old Dobbin, and ate gas instead of hay. But the general-purpose tractor, with its power take-off, planted, cultivated and harvested. It practically wiped out the 'doubtful' and 'non-tractor' operations listed in 1921.
—*Prairie Farmer*, 1941

Ford-Ferguson 9N
This Ford publicity photo pictured a 9N with its wheel removed to show details of the three-point hitch. (From the collections of Henry Ford Museum & Greenfield Village)

daily basis, hastening the decision-making process. No formal meetings were held, but the results of their conversations were passed on to Brock for incorporation into the design.

The new tractor was designated the 9N—9 for the year 1939, N being the Ford designation for tractors. The Ferguson-Brown tractor with Ferguson's implement hitch had proved to be an unsatisfactory design and was given only casual observation by Ford engineers, according to Brock.

All of Ford's resources were devoted to the new tractor program, which resulted in a prototype by early spring and the first production tractor by mid-1939—an astonishing six-month development period, according to Brock.

Ford's goal was to create an inexpensive tractor. He established a retail price of less than $600, speculating that this price would create sufficient demand to meet his production goal of 1,000 units weekly, which would be enough to keep an assembly line running.

The resulting tractor design was unique. It did not incorporate any parts from the Ferguson-Brown; Brock says parts from Ford car and truck lines were adopted for the 9N to meet Ford's price goals. The magneto system had been one of the Fordson's major problem areas, so the 9N boasted a tried-and-true Ford automotive electrical system with a coil ignition system. A typical car battery was used although the starter and generator were new for the tractor. A modified car clutch and car front wheels as well as truck differential gears and brakes were incorporated.

The engine of the 9N was an inline L-head four-cylinder. Displacement was 119.7 ci (1,960 cc) from a 3.187x3.75-inch (79.6x93.75-mm) bore and stroke. The engine was based on a single cylinder bank of the 239-ci (3,915-cc) V-8 Mercury/Ford truck engine. The 9N shared numerous components with this engine, including the V-8's pistons, rings, valves, connecting rods, and con rod pins as well as sundry bearings and gaskets. The compression ratio was 6.0:1 running on gasoline.

Whereas the V-8 was rated at 95 hp at 3,600 rpm, the gasoline tractor engine was rated at a conservative 1,400 rpm to start with, but its governor allowed operation up to 2,200 rpm. At 2,200 rpm, 28 hp was produced at the PTO. Ford later used the 9N engine in the original Ford-built U.S. Army Jeep rated at 45 hp and in light trucks rated at 30 hp.

Most 9Ns for the North American market were built to run on gasoline. The 9NAN variation was also offered, which used distillate fed through the same Holley Vaporizer used on the Fordson F. The compression ratio for the 9NAN was lowered to 4.75:1, resulting in a loss of about 10 percent of the power, down to a rating of 21 hp. Most 9NANs were exported, many to Great Britain.

The 9N retained the unified chassis of its predecessors and did not have a traditional frame. Despite the cost constraints, operator safety dictated that fenders be standard equipment, as well as a PTO and reverse-flow muffler. Running boards, however, were not included as standard until the 8N appeared in 1947.

The novel tractor incorporated what became known as the "utility tractor front axle." This straight axle pivoted in the center and had king-pins that extended downward. With this arrangement, the tractor had more crop clearance and a higher roll center for greater stability. Ford engineers also created a method for front-axle width adjustment using three beams. The outer two beams, to which the wheels were attached, were bolted to the center beam, which supported the tractor. The center beam angle allowed the outboard beams to move in or out without changing the steering geometry.

Finally, the new 9N was dressed up in stylish sheet metal crafted by the Ford styling department. Styling was a hot commodity for tractors in the late 1930s. Deere had hired New York City industrial designer Henry Dreyfuss in 1937 to streamline the bodywork on its line of tractors; International Harvester countered in 1939 by bringing in automotive stylist Raymond Loewy to give its tractors a facelift. Ford's 9N was a masterpiece of form and function. The result of the styling department's handiwork was a modern blending of practicality and art deco flair. In fact, the 9N's styling was so good that Ford changed it only slightly over the following thirteen years.

The 9N was conceived to be available with a complete system of implements. Ferguson and the Sherman brothers launched a new marketing company called Ferguson-Sherman Manufacturing Corporation to design and manufacture this line of implements. (Ferguson would later

1940 Ford-Ferguson 9N

This late-1940 9N has an aluminum horizontal-bar grille but the later steel hood. This tractor was made up of available parts near the time of transition from 1940 to 1941 models. Supposedly, Henry Ford himself came down to the assembly line when production was held up due to parts availability. He ordered a search of the spare and obsolete parts bins in order to get things moving again. This tractor has a safety starter (introduced in mid-1940), and the ignition key on the steering column like the 1941 models. It also has the hinged battery cover and single-ribbed fenders. Owners: Ken and Margaret Ellis of Rockville, Indiana. Tractor remanufactured by N-Complete.

1940 and 1942 Ford-Ferguson 9Ns

Although the steel-wheeled 9N in the foreground was built in early 1942, it has the solid center-bar grille of the 1941 model. Behind it is a 1940 9N, which has the ignition key on the left side of the dash and the hinged battery cover (the knob just ahead of the steering wheel). Emerging from the garage is a 1930 Model A Ford car. The 1942 9N is owned by Charles Hardesty and the 1940 by Louis Norfleet, both of Valparaiso, Indiana.

have a falling out with the Shermans and would start a new company, Harry Ferguson, Inc., to handle distribution of Ford-Ferguson tractors and implements.)

But as the day of the tractor's debut neared, Ferguson-Sherman had only a British plow, cultivator, and middle-buster ready for service. These were not satisfactory for North American farms, so Ford's design team quickly created the necessary implements, including a disc harrow. A two-bottom plow was shaped by the Ford car-body group copying an old Fordson plow built by the Oliver Chilled Plow Company of South Bend, Indiana. A planter was offered, although it was a straightforward copy of the famed Covington planter.

Debut of the Ford 9N Tractor

"It was a bad day for old Dobbin," read the *Chicago Journal of Commerce*'s article heralding the first public demonstration of Ford and Ferguson's revolutionary 9N. The new tractor was first shown to the new Ferguson-Sherman dealer organization in early June 1939. On June 29, a second, much grander demonstration was organized in typical Ferguson fashion with five hundred invited guests, including journalists from all over the world along with other agricultural dignitaries. Tents were erected and a full lunch served. Then, the star of the day was unveiled. Ford, Ferguson, and others—including eight-year-old David McLaren, a Greenfield Village, Michigan, schoolboy—gave plowing demonstrations at the wheel of the 9N. The press was astonished when the schoolboy's furrows were just as even as those run by experienced plowmen.

The Ford-Ferguson 9N was an immediate success. Its targeted competition was the seventeen million horses still at work on North American farms in 1939, rather than the numerous other farm tractors. Even with the requirement of buying all-new implements, the 9N was still less expensive than other tractors of similar work capability. For example, if you were a Big Green fan, you needed one of John Deere's new Model G tractors to equal the acre-per-hour plowing rate of the 9N—and a 1939 Deere G cost more than twice the Ford's price of $585.

That launch price of $585 included rubber tires, an electrical system with starter, fenders, PTO, oil-bath air cleaner, oil filter, and an automobile-type reverse flow muffler, which made the tractor as "quiet as a dragonfly." Lights were always optional.

Initially, there was resistance to the rear-mounted cultivator of the 9N, and most farmers doing serious cultivating opted for a row-crop tractor and front cultivator such as the Farmall. But the 9N's rear-mounting system would go on to set a standard, and today, almost all cultivation is done by rear-mounted units following the Ferguson concept.

The only serious weakness in the 9N was that it was too lightweight to do serious farmwork without the weight-transferring aid of the three-point implements. Thus, in its first University of Nebraska Tractor Test (test number 339), the 9N made a poor showing. Ford sent the 9N to its Nebraska Test with only 300 lb (135 kg) of calcium chloride ballast, as Ford engineers did not understand the common practice of weighting

The horse and man made civilization: they should forever stand inseparable.
—Horse Association of America's *1924 Annual Report* warning against the rise of farm tractors

1940 Ford-Ferguson 9N
Facing page, top: *This late-1940 9N was built with the solid-bar grille of the 1941 model; it was not uncommon with Ford tractors that characteristics overlapped at model-year end. Owners: Jim and Jane Woehrman of Camden Point, Missouri. Tractor remanufactured by N-Complete.*

1941 Ford-Ferguson 9N
Facing page, bottom: *This incredible 1941 9N is in original, unrestored condition. The 119.7-ci (1,961-cc) four-cylinder, L-head engine was used in Ford tractors with little change until 1950, when a side-mounted distributor replaced the front-mounted type. Owner: Darrell Craycraft of Indiana.*

1945 Ford-Ferguson 2N

Above: *This 2N has several interesting accessories. The devices attached to the rear wheels are called "sand wheels" and provide extra push in soft going. The tires are the original Firestones. The tractor has a Monroe Easy-Ride seat and is equipped with a safety clutch that prevents plow damage. Ford-Fergusons for 1945 were little different from the 1944 model, except for a more beefy rear-axle housing. Owner: Dwight Emstrom of Galesburg, Illinois.*

1945 Ford-Ferguson 2N

Right: *The business end of the Ford-Ferguson tractor. The elements of Ferguson's three-point hitch are clearly shown. This tractor has a drawbar between the two lower hitch points. To stabilize the drawbar, adjustable sway bars have been connected to the lower hitch points and extend up to the third point below the seat. Just ahead of this third point is the draft-load-compensating coil spring. For normal implement operation, the implement is connected to the three points, replacing the drawbar and sway braces. Draft loads operating on the implement react against the spring. When the spring is compressed by increasing draft loads, the implement is automatically raised. When the hard spot is passed, the spring returns the implement to its original depth.*

1942 Ford-Ferguson 9N

Above: *This 9N is equipped with steel wheels, magneto ignition, and a hand crank. No battery, starter, or generator was provided, making this 9N much the same as the initial 2Ns. The change of designations helped Ford get a price increase past the wartime price control board. This 9N is serial number 94540; the switch to the 2N occurred at number 99003. This 9N is equipped with a Sherman Steering Shock Absorber that raises the steering wheel 3 inches (7.5 cm). This bi-directional no-back device prevents front wheel forces from feeding back to the steering wheel.*

Left: **1940s French Ford-Ferguson ad**

Above: **1944 Ford-Ferguson 2N ad**

1945 Ford-Ferguson 2N

Left: *This 2N is mounted with a Ferguson two-bottom disk plow. Disk plows are used to prevent soil compaction; regular plows tend to press down on the base of the furrow leaving it smooth and hard. This made the soil beneath the tillage impervious to rainfall and enhanced the likelihood of erosion. A disk plow scooped out the furrow throwing it over and pulverizing it, leaving the base of the furrow soft and ready to transfer moisture either up or down. Owner: Dean Simmons of Frederickstown, Ohio.*

test tractors for all they were worth. To curb wheel-spin, Ford called for an engine speed of only 1,400 rpm; thus, tests showed the 9N developing only 12.8 hp on the drawbar. On the other hand, belt power tests were done at 2,000 rpm and produced a respectable 23.07 hp.

The 9N was in production for six months in 1939; the model year actually ended in November. More than 10,000 Ford-Fergusons were sold in that first half-year despite normal assembly-line dilemmas and problems with sub-contractor parts supplies.

The Simplified 9N: Introducing the Ford-Ferguson 2N

During the early days of World War II, Ford was faced with shortages of rubber and copper that affected 9N production, and so a simplified version was introduced. It had none of the amenities of a starter, generator, or lights, and was equipped with steel wheels. Reflecting the changes, the model designation was changed from 9N to 2N, for 1942. A new model number was also a way of getting a price increase past the U.S. War Production Board.

Except for minor annual changes, the 2N was much the same as the 9N. By 1943, the starter, generator, and lights, along with the rubber tires, found their way back onto the new tractor model.

The Model 2N was produced until mid-1947. With Henry Ford II at the controls of the Ford Motor Company, the Handshake Agreement would be called off, and Ford and Ferguson would part, ending one of the most amazing business and personal relationships in agriculture. The Ford-Ferguson had revolutionized farming. It had also given Ferguson a much-deserved and prominent place in agricultural history. It was truly the farm tractor of the century.

1946 Ford-Ferguson 2N
Above: *This 2N has a Sherman Step-Up transmission giving six forward and two reverse speeds. The white barn behind the tractor was built in 1910. Owner: Floyd Dominique of Napoleon, Ohio.*

1946 Ford-Ferguson 2N
Facing page: *Headlights and grill guard were options when this 1946 2N was new. The 2N enjoyed a six-year production run. Even though the first and last years were only partial years, almost 200,000 were made, more than twice the number of 9Ns.*

1946 Ford-Ferguson 2N

Above: *This 2N is equipped with the Sherman auxiliary two-speed gearbox. Although painted like an 8N, this 2N has some significant visual differences from the later model: both the front and back wheels are different (8N wheels have small-center hubs), and the steering wheel is lower on the 2N. Owner: Doug Marcum of Mt. Comfort, Indiana.*

1940s BNO-40 aircraft tug

Right: *This aircraft tug was built from the Ford-Ferguson 2N for the U.S. Navy during World War II by Ferguson-Sherman Manufacturing. The heavy sheet metal raised the weight to 4,000 lb (1,800 kg). The BNO-40 was capable of towing the Navy's heaviest World War II aircraft. It has truck-type dual rear wheels, a single brake pedal, and hydraulic brakes. Owner: Dwight Emstrom.*

1946 Ford-Ferguson 2N

Almost 60,000 2Ns were built along with this one in 1946, the year wartime shortages and restrictions were finally overcome. The end of the Ford-Ferguson Handshake Agreement was looming, however. Ferguson was told in September that the agreement would end in mid-1947, and that Ford would build the 8N without him after that.

Chapter 5

The British Fordson Farms On

Make Hay the Fordson Way.
—Fordson ad, 1926

Fordson New Major nose medallion

1959 Fordson Dexta
The Dexta was an advanced tractor for its day with its three-cylinder Perkins diesel engine. It also featured live hydraulics and a live PTO, as the badge in the grille states. Owner: Jonathan Philip of Looe, Cornwall, Great Britain.

With another war looming in 1938, Patrick Hennessey, British Ford Motor Company general manager, approached the British Ministry of Agriculture and Fisheries. To prevent another tillage crisis, he reasoned, a quantity of Fordsons could be built ahead of time and stored ready for use if war came. The plan allowed the Ministry to buy 3,000 Fordsons at a reduced price, and the tractors would be distributed to dealers throughout Britain and kept in readiness. If war did not come, the dealers could sell the Fordsons, and the government's money would be refunded. It took an act of Parliament, but the plan was approved in May 1939.

War came in August 1939, and the 3,000 Fordsons were ready for service. They were soon joined by thousands more from the Dagenham factory until production was stopped at the war's end in 1945. Their contribution to food production was significant, a factor that endeared the little machines to the British farmers from then on.

Australian Fordson Major ad

Unaware of the Fordson agreement, Ferguson sailed for Britain in September 1939. The U.S. operations were left in the hands of the Sherman brothers. Ferguson had arranged for a Ford-Ferguson tractor and implements to be shipped to Belfast. He then staged several demonstrations comparing the Ford-Ferguson to competing tractors, to the unanimous amazement of government and academic officials, and officials of the Ford Motor Company. His real objective was to compel Ford to switch from the venerable Fordson to the new 9N Ford-Ferguson. He had also, in the name of Henry Ford, demanded a seat on the British Ford Motor Company board of directors.

Lord Percival Perry, Chairman of Ford Motor Company (England) Ltd., was not about to get involved with the feisty commoner who had the reputation of demanding his own way. The other directors felt the same way. To them, he was a presumptuous upstart from the Irish bogs. These feel ings of the traditionally upper-crust board members were not without foundation. It was said that during one of the 9N demonstrations, Ferguson was infuriated by war planes flying over on their way to assist the embattled troops at Dunkirk. He had demanded that a Ford public relations official call the Royal Air Force to have the traffic detoured.

With Britain now at war, the government was dictating the expenditure of resources. Any change to the current-production Fordson was out of the question. Ferguson reluctantly decided to bide his time until the end of the war.

The Great Fordson Major E27N
When the end of World War II was on the horizon in 1945, the British War Agricultural Committee set the specifications for a new farm trac-

1946 Fordson E27N Major

The Fordson Major made its debut following the end of World War II. While it was firmly based on the prewar Fordson Model N, it also featured many revised, updated, and uprated features. Owner: Ivan Sparks of Street, Somerset, Great Britain.

tor; the committee allocated materials, and if a maker did not conform to the specifications, it did not receive materials. British Ford Motor Company was interested since there would undoubtedly be a considerable market. The new specifications called for three-plow capability, central PTO, and higher crop clearance (or row-crop configuration). British Ford, still fearful of a relationship with Ferguson, saw the three-plow requirement as grounds for eliminating the "gray menace," the Ford-Ferguson 2N, from consideration.

The building of about 150,000 Fordson Model Ns during the six wartime years had severely taxed the production facilities. When the war ended, there was no time, material, or viable machine tools for a completely new tractor design. Therefore, British Ford undertook a complete upgrade and modernization of the venerable Fordson, and U.S. Ford engineer Harold Brock went to Dagenham in 1946 to help design the updated tractor. It was to be designated the E27N—E for English, 27 for the horsepower of the Fordson engine, and N remaining the Ford designator for tractors. The commonly used name for the E27N was the Fordson Major. The E27N designation was not much used until it became necessary to differentiate between the E27N and later Ford tractors, also called Majors.

True row-crop tractors, with tricycle-type front wheels, were never as popular in the United Kingdom as they had been in North America. The need for more crop clearance than that provided by the original

Fordson was a requirement, however, so the new version would have to be built higher. Brock designed the tractor for higher crop clearance, a live PTO, improved styling, and better controls. Some of these features were previously included in an experimental 4P model developed at Ford in the mid-1940s, according to Brock.

A redesigned rear axle with a spiral bevel drive and bull gears was incorporated since the original worm-drive rear end would not take the additional load of pulling three plows. The new rear axle, combined with a front axle with downward-extending kingpins, gave the new E27N the desired higher stance. The higher tractor would no longer fit through the tractor assembly-line paint booth, however, so E27Ns were at first painted by hand at the end of the truck assembly line. The color reverted to the blue-and-orange scheme of 1938.

1946 Fordson E27N Major
The three-plow capability of the E27N severely taxed the uprated original Fordson engine. Rebuilding of the engine was an in-shop project, due to the poured bearings.

The Fordson's engine design was retained although it was now almost thirty years old. Displacement of the new engine was increased from the old Fordson's 251 ci (4,111 cc) to 267 ci (4,373 cc) with a 4.125x5.00-inch (103x125-mm) bore and stroke. The four-cylinder L-head engine fathered its 27 hp at 1,200 rpm. The old engine was barely adequate, especially the high-compression gasoline, or petrol, version. Worse, overhaul was an in-shop project, since the cylinders did not have sleeves and the main bearings were made of babbitt. To keep the E27N viable, an engine exchange program was developed.

Help for the E27N was on its way. In 1950, Frank Perkins of F. Perkins, Ltd., of Peterborough, England, converted his own Fordson E27N to a Perkins diesel engine. The conversion worked so well that when British Ford heard of Perkins's work, it sent two more E27Ns to Perkins for modification. The six-cylinder E27N P6(TA) Fordson diesel was the result. With a bore and stroke of 3.50x5.00 inches (87.5x125 mm), its 288-ci (4,717-cc) engine was rated at 45 hp at 1,500 rpm. Before production of the P6(TA) ended in 1952, some 23,000 six-cylinder diesels were built.

Perkins also created a four-cylinder P4(TA) engine for the E27N in 1953. The P4 conversion kits were available to change either the P6 or P4 into E27Ns.

The E27N used two types of hydraulic lifts, one made by Smith and the other by Varley. The three-point linkage was originally connected to the lift arms by chains; later, solid links were used with a leveler. The Varley lift used two control levers, the Smith lift used one. The Hydraulic Power Lift (HPL) was available after early 1948. The system did not provide draft control.

Henry Ford II and Fordson E27N
Henry Ford II tries his hand with a new E27N Fordson Major at the Dagenham, England, plant. (From the collections of Henry Ford Museum & Greenfield Village)

There were several versions of the E27N. The Standard Agricultural Model had 9-inch-wide (22.5-cm) steel rear wheels, a single brake, and fixed wheel tread. The Row Crop version had 4.5-inch-wide (11.25-cm) steel rear wheels, individual brakes for each rear wheel, and adjustable wheel tread. The Land Utility type, or LU, was like the Row Crop, but factory equipped with rubber tires. Ford also created industrial versions, and crawlers were developed in association with the companies County Commercial Cars of Fleet, and Roadless Traction, Ltd., of Hounslow, Middlesex.

Debut of the Fordson New Major E1A

The major impact the Fordson E27N Major had on the tractor world was the popularizing of diesel power. Except for the highly successful Farmall MD, diesels had been viewed with suspicion; farmers thought them too costly and temperamental for routine use. But after a few years of the Farmall MD and the Fordson Perkins Major, distillate tractor fuel was on its way out.

Dagenham engineers had foreseen diesel as the wave of the future even before Frank Perkins had installed his diesel engine in his E27N. In

1946 Fordson E27N Major

Above: *The spark-ignition gasoline-TVO engine of a 1946 E27N Fordson. The Major's engine was increased in size and power from the old Fordson's 251 ci (4,111 cc) to 267 ci (4,373 cc). The four-cylinder L-head engine created 27 hp at 1,200 rpm.*

1958 Fordson New Major

Facing page: *The first all-new tractor from British Ford was the New Major, introduced in 1952 to replace the aging E27N Major. Most New Majors are now simply called Fordson Majors, and E27N Majors are called E27Ns. This late-1958 New Major has the rare narrow-front configuration. Owner: Dale Bissen of Adams, Minnesota.*

1944, British Ford General Manager Patrick Hennessey authorized the design of a new tractor that would use an engine offered in diesel as well as in gasoline and TVO (Tractor Vaporizing Oil) versions. In the end, however, the E27N was a stopgap measure due to material restrictions and development delays in the postwar years. The "New" Major, as it was called, was not ready until 1952.

The same engine block and crankshaft were used in all three of the New Major's engine variations. The 199-ci (3,260-cc) engine had bore and stroke of 3.74x4.52 inches (93.5x113 mm). Compression ratios ranged from 4.35:1 for TVO to 5.5:1 for gasoline to 16.0:1 for the diesel. Diesel versions had a larger bore, and thus a displacement advantage over the gasoline and TVO versions. The diesel New Major was rated at 45 brake hp, about the same as the E27N with the six-cylinder Perkins P6 engine. By 1960, more than 90 percent of production was diesel powered.

The new E1A tractor was larger and heavier than the E27N but shared the same general layout. The hydraulic system differed from the E27N, being more like that of the Ford-Ferguson. The linkage was much the same as that of the E27N. Draft control was not offered at first, but the live PTO option was added in February 1957. The E1A used the same three-speed transmission as the E27N, but with a two-range auxiliary, giving it six forward and two reverse speeds. A true tricycle row-crop option was provided for the North American market throughout production.

1952 Fordson New Major

Above: *When the New Major made its debut in 1952 it was a departure from the usual Ford tractor design in that it featured a front frame, rather than relying upon the castings for structure. Owner: Duke Potter of Chippenham, Wiltshire, Great Britain.*

Right: **Australian Fordson New Major ad**

The New Major departed from the unit-frame concept that the original Fordson pioneered in volume production. Instead, frame channels extended from the clutch housing around in front of the radiator.

The Fordson Power Major and Super Major

In July 1958, the Power Major was introduced. The Power Major, and subsequent Super Major, featured a 220-ci (3,603-cc) engine with bore and stroke of 3.94x4.52 inches (98.5x113 mm). The TVO engine option had been quietly dropped during the last days of the New Major, and the diesel fuel-injection system improved to increase power dramatically. Whereas the New Major had been rated at 45 brake hp, the Power Major bragged of a diesel engine with 52 bhp at 1,600 rpm. In 1962, a change to the Simms Minimec injector system raised power to 54 bhp.

The Super Major made its debut in October 1960, as an upgraded Power Major with added features. Ferguson's draft control was added to the hydraulics as his patents had lapsed in England.

Disk brakes were added to the Super Major, as was a manual differential lock. The headlights were repositioned to the grille panels to eliminate interference with front-end loaders, which were becoming increasingly popular. In 1962, as Ford began consolidating the tractor manufacturing operations, the Super Major was exported to the United States in the new blue-and-cream livery and labeled the Ford 5000 Diesel.

1952 Fordson New Major
Duke Potter sits proudly on his masterpiece of restoration handiwork. This tractor has a PTO, hydraulics, and the spark-ignition engine variant with "underneath" exhaust.

1962 Fordson Super Major

Above: *The Super Major came out in fall 1960. It offered many improvements over the previous model, the Power Major, such as disk brakes, differential lock, and a draft-control three-point hitch. Owner: Jonathan Philip.*

1958 Fordson Power Major

Left: *The Power Major was built between 1958 and 1961. It used a Ford-built four-cylinder diesel engine of 220 ci (3,604 cc). At 1,700 rpm, its six-speed transmission gave speeds of less than 2 mph to almost 14 mph (3.2 to 22.4 km/h). During University of Nebraska Tests in 1959, the Power Major demonstrated a maximum drawbar pull of over 8,000 lb (3,600 kg)—78 percent of its own weight. Owner: Carleton Sather of Northfield, Minnesota.*

No development in the industry was regarded with more distrust and wholesale opposition then the suggested general purpose tractor.
—Agricultural Engineering, 1931

In June 1963, the New Performance Super Major replaced the Super Major. Engine improvements brought power up to 54 hp. The paint scheme was changed from the familiar blue and red to blue and gray. Hydraulics were improved, and transmission ratios were modified to optimize PTO work.

Enter the Fordson Dexta

British Ford finally unveiled its version of the Ford-Ferguson in 1957. Harry Ferguson was not involved, of course—except as competition. The new Fordson Dexta had all the Ferguson System features, including the draft-control three-point hitch. It was British Ford's first compact tractor since the Fordson N.

When the Fergie was introduced to the British farmers after World War II, it proved to be stiff competition for British Ford's new E27N. Then, in the early 1950s, Ford's marketing people saw what they believed to be a golden opportunity in the larger tractor sector. The development of the E1A New Major followed, consuming the lion's share of Ford's resources. By 1955, work had begun on a scaled-down version of the New Major that would eventually become known as the Dexta. Ford considered a three-cylinder version of the E1A engine but dropped the idea since it did not have the manufacturing capacity for another engine line.

Ford engineers then turned to Frank Perkins once again. The Perkins P6 diesel had been adapted to the E27N and the Ford Thames truck with considerable success. Perkins P3 three-cylinder engines had been used as replacements in the few 9Ns and 8Ns imported from the U.S., as well as in the tractors Ferguson built after his falling out with Ford. However,

1962 Fordson Super Major
The Super Major was a competitive tractor in the 50–60-hp range. With its six forward and two reverse speeds, it was suited for a variety of tasks. This 1962 Super Major has about 2,000 hours of experience. When Dean Simmons bought it, it was in bad shape and inoperable. He changed the fuel filter, and it started right up.

this P3 engine was not available in time for use as a production diesel in the Ferguson TE-20 tractor, so Ferguson employed the Freeman Sanders diesel design. The Perkins P3 was just right for the British equivalent of the 8N/NAA, the Dexta. It provided great low-end torque, low fuel consumption, and reliable cold-weather starting. It is interesting to note that throughout the production life of the Dexta, Ford cast the blocks and cylinder heads; these rough castings were then trucked to Perkins, and finished engines made the return trip.

The Dexta had 32 hp on tap thanks to its Perkins diesel engine. The Perkins was a 144-ci (2,359-cc) three-cylinder unit rated to 2,000 rpm. Power was fed through a three-speed transmission and a two-speed auxiliary, resulting in six forward and two reverses speeds, similar to the E1A New Major and the later Ferguson TO-35. At 2,000 rpm, the Dexta had a first-gear speed of only 1.72 mph (2.75 km/h) and the high-gear top speed was 17 mph (27 km/h)!

A non-diesel Dexta was offered in limited quantities. The gasoline, or petrol, version used the same Standard Vanguard car engine that would be used in the Ferguson TE-20 tractor, except that it was bored out to 3.43 inches (8.5 cm). This four-cylinder gas engine was longer than the Perkins, so the hood, steering rods, and radius rods had to be extended 4.3 inches (10.75 cm).

Originally, the Dexta had an optional live PTO with a two-stage clutch. Optional live hydraulics were also offered that could accommodate two

1962 Fordson Super Major
The Super Major incorporated the Ferguson draft-control three-point hitch. The basic weight of the Super Major was about 5,500 lb (2,475 kg), but ballast commonly brought it up to 8,000 lb (3,600 kg) or more. British E1A Fordsons' bodies were painted blue with orange wheels. When imported to North America as the Ford 5000 Diesel, a blue-and-cream scheme was used.

1959 Fordson Dexta

Above: *The Fordson Dexta was built from 1957 to 1964. The Dexta used a Perkins P3 engine as Ford did not have the production capacity to manufacture another engine. Ford did, however, cast the blocks and deliver them to Perkins for finishing.*

1958 Fordson Dexta

Right: *When introduced in 1958, the Dexta from Dagenham, England, rocked the competition—including Ford of Detroit. British Ford out-produced U.S. Ford by 58,000 to 46,000 tractors respectively in the Dexta's debut year. Owner: Carleton Sather.*

remote cylinders. Non-live PTO and hydraulics were standard, but not often selected. The 3,000-lb (1,350-kg) Dexta was rated a three-plow tractor, as was the E27N. The TE-20 was rated for two.

The original Dexta had a silhouette similar to the North American N Series tractors. In 1960, the Dexta was restyled with the headlights placed in the grille like those of the new Super Major.

The Super Dexta was introduced in 1962. It was essentially the same as the original Dexta, except it was powered by a larger, 152.7-ci (2,501-cc) three-cylinder diesel with a 2,200 rpm rated speed, producing approximately 39 hp. The headlights were retained in the grille to prevent interference with front-end loaders. Another feature of the Super Dexta was a differential lock to improve traction. In 1962, Super Dextas were exported to North America as the Ford 2000 Diesel.

The New Performance Super Dexta debuted in 1963. Rated engine speed was increased to 2,450 rpm and output increased to 45 hp.

Production of the Dexta ended in 1964, after sales of about 150,000 units worldwide. The paint scheme was a medium blue with red wheels. For those sold in North America as the Ford 2000 Diesel, the paint scheme was the same medium-blue body but with beige wheels.

Above: **1962 Fordson Super Major Doe Triple D**

1961 Fordson Dexta
Facing page: *The British-built Fordson Dexta used a three-cylinder Perkins diesel engine. This Dexta is equipped with a side-delivery rotary hay rake. The Dexta was the first postwar compact tractor design by British Ford. The name "Dexta" is derived from the word "dexterity," meaning "handy." Owner: Dean Simmons.*

1964 Fordson New Performance Super Dexta
Above: *The New Performance version of the Dexta first appeared in 1963. It had a power increase over previous Super Dextas. Owner: Jonathan Philip.*

Fordson 5000 Diesel
Right: *The 5000 Diesel Super Major offered a Ford-built 220-ci (3,604-cc) four-cylinder diesel engine with overhead valves. The transmission provided had three speeds with two ranges in each gear. There were double disk brakes and a differential lock. This one, now owned by Fred Bissen, was once owned by Deere & Company for test purposes.*

The light tractor has made tractor farming possible for countless numbers of small farmers.
—*Capper's Farmer, 1935*

Chapter 6

Rebirth of the Ford Tractor

This company is not dying, it's already dead.
—Jack Davis, Ford executive, 1945

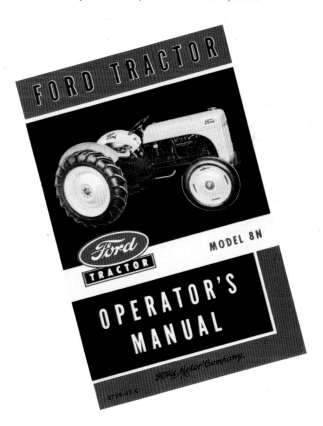

Above: **1952 Ford 8N Operator's Manual**

1949 Ford 8N
Left: *Following the break from Ferguson, Ford produced the 8N tractor from 1947 to 1952. Owner: Floyd Dominique.*

When Henry Ford's only son, Edsel, died in 1943, a serious vacuum was left at the top of the Ford Motor Company. Harry Bennett and Charles Sorensen were the more-or-less obvious heirs apparent. The U.S. War Production Board was especially concerned, since the Willow Run plant was still having trouble getting up to speed producing B-24 Liberator heavy bombers for the war effort.

When old Henry Ford was asked whom he favored, he indicated that Harry Bennett was his choice. Bennett, the former bar bouncer, union basher, ersatz thug, and braggart that had Michigan's politicians and underworld in his pocket, was head of the Ford Service Department. Bennett was also Ford's alter ego and confidant.

Edsel's widow, Eleanor, held 41 percent of the family-owned company stock, and she was well aware of the conflicts between Edsel and Bennett. Edsel Bryant Ford had married Eleanor Lowthian Clay in 1916. While Eleanor was never bossy or overbearing, she exerted considerable influence in helping her husband avoid his father's dominance. Eleanor was also the mother of Henry Ford II, and brought him up to be his own man. Both Eleanor and Henry's wife, Clara, were adamant that the company should remain in the family's hands. Their choice to head the company was twenty-five-year-old Henry II, then in the Navy completing basic training. It was obvious, however, that he was not ready for such responsibilities. Henry II was given a special honorable discharge by the Secretary of the Navy so he could return to the company and get involved in the management situation that was developing.

When the board met, Henry himself, now eighty years old and slightly senile from a series of minor strokes, was named president. Bennett continued to run the company, as he was the only one who had old Henry's ear. Bennett also sometimes declared what Henry had said whether he had actually said it or not. Within a year Sorensen resigned, not willing to take the continuing assault from Bennett.

This situation continued until the end of World War II. During the summer of 1945, Clara finally prevailed upon old Henry to relinquish control, not to Bennett, but to her and Eleanor's choice, Henry II. Henry II took over in September 1945. His first official act was to fire Bennett.

Henry Ford would live another eighteen months, spending much of his time at Greenfield Village. The automobile magnate died April 7, 1947, at the age of eighty-three.

On the night Henry died, there was a tremendous rainstorm in the

Spy photo

Deere & Company obviously kept close tabs on Ford. In this John Deere comparison photograph, a 1952 Ford 8N is shown next to a Deere experimental tractor. The Deere machine even rides on Ford wheels. (Deere & Company)

1948 Ford 8N

Although production began in mid-1947, those models are considered early 1948s, since the 8N designation signifies 1948 (the 9N was introduced in 1939 and the 2N in 1942). This 8N is better than new after it was completely remanufactured by N-Complete. Driving one of these "new" tractors is a real treat. The steering is tight; the clutch smooth and predictable; the brakes are tight, effective, and even; the throttle linkage and governor are responsive; and the engine sound is a subdued purr.

N-Complete: New Life for Old Fords

Ford tractors manufactured between 1939 and 1952 are now being remanufactured by the firm N-Complete. Owner Tom Armstrong started the business in 1988. Since then he has given new life to hundreds of old Fords.

Armstrong starts with a Model 9N, 2N, or 8N Ford tractor in virtually any condition. The only requirement is that the castings must be sound. From there tractors are stripped to their basic parts, including engines, generators, gearboxes, differentials, axles, brakes, and hydraulic components. Radiators are recored or replaced. Tires are replaced. Good fuel tanks are pebble tumbled, and coated. Unless perfect, sheet metal is replaced. N-Complete has stamping facilities to make new hoods and fenders, and has been authorized to emboss the Ford logo as needed.

The tractors are then rebuilt around the reconditioned castings. Every component is returned to new specifications—finishes and dimensions that are as good or better than as originally built. In most cases, gears can be reused and are kept together as a set. When the bearings are renewed, the gears go back to their original wear pattern. N-Complete employs a team of expert finishers. The paint is one thing that's better than new. They also make and install their own decals.

Most of the "new" tractors Armstrong sells are used as "estate tractors." These are generally returned to the same configuration that they were when first built. Absolute correctness is not a stipulation, however, unless requested by the buyer. Unless otherwise stipulated, tractors will have modern-tread tires, and may have aftermarket rear wheels and/or lights. Collectors who are concerned for minute correctness details are also accommodated, but it may increase the cost. Armstrong has remanufactured 9Ns with aluminum hoods, as well as Funk sixes and eights, bringing them up to 100 percent historical correctness.

If the tractor to be remanufactured is owner supplied, the original engine block with the serial number is reused if possible. Accessories such as wheel weights, special seats, and so on are returned to the tractor when completed.

Armstrong got into remanufacturing Fords by accident. He had five acres (2 hectares) of stubble and brush to keep mowed. So-called garden tractors cost an arm and leg, and didn't seem to have the beef for the tough job. Armstrong bought a Ford and disassembled it down to gears and bearings. Being an engineer himself, he was impressed by the renewability of the design and by its basic ruggedness. Although it was originally one of the less-expensive tractors, Armstrong found it to be a high-quality piece of equipment. When the tractor was restored, friends and neighbors began requesting the same rebuild.

N-Complete now has a backlog of orders. About half of the customers supply the tractor to be remanufactured. Armstrong estimates there are still a half million Fords out there waiting to come to N-Complete. He thinks the tractors are still so popular because they are user friendly and well engineered. They are light, safe, have a low center of gravity, and are easy to operate. They don't have a lot of bells and whistles. They have a straight-forward manual transmission and three-point hitch. They are ergonomically well designed and are economical to maintain and run. Most importantly, remanufactured by N-Complete, they are less than half the price of a new, similar-sized tractor. N-Complete tractors carry a one-year new tractor guarantee.

As for power, Armstrong was challenged by the owner of a Japanese-built, diesel, four-wheel-drive tractor to a "pull." After being dragged around backwards by Tom's 8N for a while, the challenger's hydrostatic transmission finally blew.

N-Complete has several area dealerships established. One interesting aspect of being a N-Complete dealer is that to get "new" tractors, you've got to find old ones to remanufacture. N-Complete is also in the parts and service business. At this writing, they are gearing up to remanufacture Ford models made between 1952 and 1964.

Detroit area, and Ford's caretaker of the Fair Lane powerplant shut down the boilers due to flooding. (Reminiscent of his old days, Ford retained his own steam-power generating station.) Fires were lit in the Fair Lane mansion fireplaces, and candles and kerosene lamps were placed about. After a tour to inspect the flooding, Ford went to bed at 9:00 P.M. He awoke at 11:15 P.M. feeling ill and thirsty. Clara, his wife of fifty-nine years, brought him a drink of water. Then, some time just before midnight, he died quietly in a room warmed by an open fire and lit only by candles. And so Henry Ford left this world in a scene much like his Michi-

1948 Ford 8N
The 8N's three-point hitch system was essentially the same as the Ferguson system used on the Ford-Ferguson 2N, which resulted in Harry Ferguson's bitter patent infringement lawsuit. This rear view of an 8N shows the fine detail work done by N-Complete.

gan farmhouse birth.

Ford was buried in a family plot a short distance from where he was born. Today, the plot is on the front lawn of St. Martha's Episcopal Church, which was built a decade after his death. Ford's fortune was estimated at well over $500 million. After his death, it was discovered that he kept a $26 million cash fund in a separate bank account for contingencies.

Henry Ford has been compared to John Rockefeller, Cornelius Vanderbilt, and J. P. Morgan, but there is an important difference. Ford did not make his money through financial manipulations, or by cornering a market. Ford produced good products that people needed and could afford. Henry Ford had more of an impact on modern man than any tycoon, politician, industrialist, military leader, or inventor, except perhaps for his friend, Thomas Edison. The Ford car was affordable to

1949 Ford 8N

The 8N engine was, at least at first, the same as that used in the 9N and 2N. The 8N used a four-speed rather than a three-speed transmission, however. The 8N styling was essentially the same as that of the 9N and 2N—and that of the Ferguson TE-20 and TO-20, for that matter. There were many evolutionary improvements, however. On the 8N, both brake pedals were on the right side and could be operated individually, or together, by the right foot. This allowed the left foot to operate the clutch. Sharp left turns from a stop were much easier to perform with the 8N. When Floyd Dominique bought this 8N for $300, it had weeds growing up through it.

almost everyone, as were the Ford tractors, built both in Detroit and in Dagenham, England. Henry Ford put the world on wheels.

Dissolution of the Handshake Agreement

World War II ended on September 2, 1945, when Japan formally surrendered. Ford Motor Company's business had been disrupted during the war by military production. June saw the last of the giant B-24 Liberators roll off the line at the Willow Run plant. A testimony to the production genius of Charles Sorensen, in two years and ten months, 8,685 of the four-engined planes were built. The factory rolled out one of the 36,000-lb (16,200-kg) bombers every hour.

Tractor operations were probably the least affected by the war, as tractor production was considered essential to the war effort. In 1945, production of the Model 2N hit 28,729 units. While this was not much for Ford compared to its other manufacturing divisions, it was a lot for the tractor industry, and was Ford's second-biggest war year.

Still, when Henry II took the company reins, he had to take immediate action to halt financial hemorrhaging. The tractor operation was revealed as never having been profitable. Henry II soon had to call for help from an unexpected quarter.

During World War II, a group of ten U.S. Army Air Corps officers working as a program-management team developed a particularly successful management style. When the war's end came in sight, they decided to remain a team and sell themselves to industry once they were mustered out of the service. In 1946, one of them saw an article in *Life* magazine describing the difficulties Ford was having adjusting to the postwar period. It wasn't long before the group presented itself to Henry II and was hired. They became known as the Whiz Kids. This group, and several other individuals hired at the same time, would have drastic and long-lasting effects on Ford's future. Some of the more famous names among them include Arjay Miller, Charles "Tex" Thornton, Robert S. McNamara, and Ernest R. Breech.

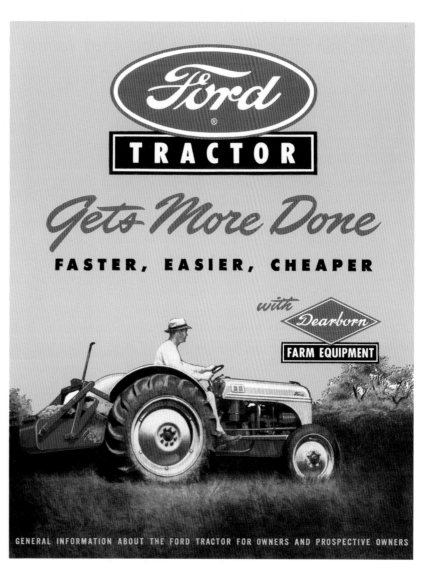

1950s Ford 8N brochure

Above: **Ford aftermarket Tractor Stilts ad**

1952 Ford 8N high-crop

Right: *This rare 8N high-crop tractor is an aftermarket modification for vegetable gardening. The tractor has 1,900 hours on it and the original 11x40-inch (27.5x100-cm) rear tires. That 8Ns would be modified to this extent, when myriad production row-crop tractors were available in 1952, is a testimony to the competitiveness of the 8N design. This tractor, owned by Dwight Emstrom, was bought from a New Jersey farmer who used it as the only tractor on his fourteen-acre (5.6-hectare) vegetable farm. Its main implement was a three-row cultivator.*

*By the dreams he had, pursued and achieved,
the burdens of drudgery were taken from the
shoulders of the humble and given to
steel and wheel.*
—Requiem for Henry Ford
by Edgar A. Guest, 1947

1948 Ford 8N

Above: *Don Artman looks as happy today on the seat of his 1948 8N as he did when he was a kid. His dad traded the family's John Deere Model D and a team of horses for an 8N so that Don could take over the farming. Don's dad had gotten a hernia hand-starting the Deere D.*

Because the Ford Motor Company stock was family held, it was difficult for Henry II to attract such talent without offering stock options. A scheme was devised by and for the new management to incorporate "sub-companies" that would offer public stock independent of Ford Motor Company stock. Executives of these companies could then profit directly from their skills through stock ownership.

One of these new sub-companies was Dearborn Motors, Inc., which was organized to distribute tractors and implements. Up until then, Ford tractors were sold to Harry Ferguson, Inc., at a fixed price; Ferguson made the implements, organized the dealer network, and made all the profits. As long as old Henry was alive, no one called the profit situation into question. The Whiz Kids could not understand it.

The arrangement with Ferguson was untenable since it left Ford with no control over the marketing of its own product. Henry II offered to buy out Ferguson in part or in whole. But Ferguson was not amenable to these overtures. He had the arrangement just the way he wanted it.

The handwriting was on the wall. Finally, on September 5, 1946, the Ford Motor Company announced that following June 30, 1947, it would discontinue manufacturing tractors for Harry Ferguson, Inc. Ford would introduce a new and improved tractor that it would distribute through its own, newly established independent dealers. The Handshake Agreement was null and void.

Henry II announced the formation of Dearborn Motors about two weeks later. Dearborn was to design and build a completely new line of proprietary implements to be marketed along with a new Ford tractor, dubbed the 8N.

Ferguson replied just as expected. In January 1947, he filed two lawsuits against the Ford Motor Company, asking for $251 million. The first suit charged Ford with seeking to create a small-tractor monopoly, resulting in making Ferguson's business unprofitable. The second suit charged Ford with infringing on Ferguson's patents. Henry II shot back, "The blunt truth is the Ford-Ferguson deal made Harry Ferguson a multimillionaire and cost Ford $25 million."

The question of who truly owned rights to design concepts and patents was always a point of battle. Ford and Ferguson engineers had created many of the Ford-Ferguson tractor design concepts jointly, or they were thought up by Ford people independently—but Ferguson obtained the patents, often with the approval of Henry Ford, according to Harold Brock. From the start, Ferguson was not a great engineer, but he was a far-seeing visionary, and he had been careful to have patents under his own name. Whereas Willie Sands had designed the three-point hitch,

Ford's broad Dearborn implement line for 1949

Ferguson had patented it under English patent law that allowed the patent to be owned by someone other than the inventor. Ferguson knew that Henry Ford put little stock in patents ever since he had so easily beaten out George Seldon's automobile patent at the turn of the century. Thus, Henry Ford granted Ferguson rights to obtain patents on the Ford-Ferguson tractor.

On the new 8N, the old front-wheel spread-adjustment idea would not be used, although what was used looked similar. The 8N also included the three-point linkage and draft control, as the Ferguson patents were running out anyway.

But Ferguson still held a trump card. His real claim was the hydraulic control scheme invented before the Handshake Agreement. With this system, flow was controlled by restricting the supply side of the pump. Ford engineers did not have time to revise the system for the 8N, although they had always considered it to be the hard way to do the job.

It was a legal battle of the titans, and it was long and fierce and expensive. In the two lawsuits, the hydraulic control was ultimately the only point on which Ferguson won. In an out-of-court settlement approved by the court, Ford was instructed to stop using supply-side hydraulic control by the end of the 1952 model year. Ferguson had meanwhile started his own tractor company, and its success belied loss-of-business damages. The settlement was for $9.25 million, a fraction of the amount Ford had spent in its legal defense.

1948 Ford 8N

Above: *The Ford 119.7-ci (1,961-cc) four-cylinder engine was rated at 21.95 hp at 2,000 rpm. The tractor was much lighter than the competition, at around 2,500 lb (1,125 kg) without ballast. The 8N sold for around $1,000 in 1948. This 1948 8N has the same front wheels as the 9N and 2N. They may have been added during its life, or since it was Ford's policy to use up leftover parts, it may have been built that way.*

1952 Ford 8N

Left: *Tim Yarberry is only the second owner of this 1952 8N, as his grandfather bought it new. Tim still has his grandfather's original canceled check paying for the tractor. Tim runs a large landscape company that uses new Ford 250C diesel tractors. Tractor remanufactured by N-Complete.*

1952 Ford 8N with Dearborn 19-35 grader

The grader was a Ford-offered option as Dearborn Model Number 19-35. The grader assembly was made for Ford by Blumberg Manufacturing Company of New Holstein, Wisconsin. An engine-driven hydraulic pump supplies blade angling and tilting cylinders. The tractor retains its three-point hitch. Owner: Dwight Emstrom.

Debut of the Ford 8N Tractor

In the early days of 1947, the successful battle against Ferguson's two lawsuits in 1952 was not a foregone conclusion. A new Ford tractor was needed to launch the new Dearborn Motors dealer organization, so Harold Brock's design team hurriedly began work. Those involved in Dearborn Motors also started work, not knowing for sure if they would have a tractor to sell.

In July 1947, the tractor and its implements were ready. Model 2N production stopped and 8N production took over. The enthusiastic reception by farmers made 1947 the best year for Ford tractor production in twenty years.

The 8N differed from the 2N in many ways, although they looked much the same from a distance. The most noticeable difference was the lighter shade of gray sheet-metal paint and the contrasting bright red cast iron of the 8N, instead of the darker all-gray paint of the older tractor.

At first, the four-cylinder L-head engine was virtually identical to the 2N, retaining the old 119.7-ci (1,960-cc) displacement with a 3.187x3.75-inch (79.6x93.75-mm) bore and stroke. The compression ratio was 6.0:1 early on, although it was later increased to 6.7:1 on gasoline. Peak power of 25.5 PTO hp came at a governed 2,000 rpm. The first 8Ns weighed in

at 2,410 lb (1,085 kg); late models weighed 2,490 lb (1,120.5 kg). A four-speed transmission replaced the old three-speed gearbox.

A major improvement for the 8N placed both left and right brake pedals on the right side, rather than one on each side as before. This allowed for operating the clutch and the left brake at the same time, which was not possible previously. The steering wheel was raised, the seat could be tipped up, and running boards were standard, allowing for operation while standing. The worm-and-sector-gear steering system of the 2N was replaced by a ball-screw type in 1949.

A modified version of the 8N, the 8NAN, was configured for operation on distillate. The compression ratio was lowered to 4.75:1, which resulted in a power decrease to about 23 hp. The 8NAN version accounted for only a few sales, however. In 1950, a change was made to the distributor, moving it out to the side from its previous location under the water pump. A conventional cylindrical coil was also used.

The excitement surrounding the updated Ford tractor forced many Ferguson dealers to change allegiances to Ford, according to Brock. Many of these dealers had previously worked with the Sherman brothers and the Fordson, and they now wanted to jump on the bandwagon with the 8N. In a short time, the new 8N constituted 25 percent of total U.S. tractor production.

1952 Ford 8N

Sales of the classic 8N averaged more than 400 tractors per day over the production run of five and a half years, giving it the highest sustained production rate of any tractor model. The attractive styling, performance, economy, and ease of operation keep most of them still in use after almost fifty years. This 8N, however, is the cornerstone of the Dwight Emstrom collection and, except for parades, is fully retired.

Chapter 7

The Hot-Rodded Funk-Ford Tractors

We were out making sales calls and happened to stop at this Ford tractor dealer in Milford, Illinois. . . .
—Joe Funk, 1996

Above: **Rare Funk conversion kit ad**

1952 Funk-Ford V-8 8N
Left: *Most Funk-Ford V-8s use the dual, vertical, straight exhaust pipes that gives the operator unbeatable stereo exhaust music. Owner: Robert Meyer of Gurnee, Illinois.*

Tractor hot-rodders

Twins Howard, left, and Joe Funk grew up with the aviation industry. Casting experience led them to build the Ford tractor six- and eight-cylinder engine conversion kits. In the 1950s, Funk-Fords were the most powerful gas wheel tractors ever made on a production basis. (Funk Brothers)

Funk Model B-85-C airplane

The original Funk B airplane was powered by a highly modified Ford Model B car engine. This Funk B-85-C, number N1654N, is powered by an 85-hp Continental, and is the last one to come off the assembly line. It was previously the Funk brothers private airplane. Owner Arlo Maxfield of Northville, Michigan, was a Ford employee building B-24 bombers at the giant Willow Run plant.

Funk-Ford: few names go together quite like these, and few are better in conjuring the image of brute farming power. Henry Ford's original adage, "A light car highly powered can go places a heavier car cannot go," applies also to tractors. And the Funk-Fords personify high-powered light tractors.

What, you ask, is a Funk-Ford? It is a Ford 9N, 2N, 8N, or NAA tractor with its original four-cylinder engine replaced by a Ford Six or V-8 thanks to an adaptation kit provided by Funk Aircraft Company of Coffeyville, Kansas. The remarkable synergy of this combination tripled the original power without radically changing the tractor's configuration, and without much of a price increase. Funk produced about 5,000 of these kits between 1951 and 1954.

It is said that failure is an orphan, while success has many fathers. That is certainly the case with the converted Funk-Ford tractors. The Funk brothers, however, do not claim actual paternity in this interesting story.

From Airplanes to Farm Tractors

Twins Joe and Howard Funk were born in 1910 in Akron, Ohio. Howard died in 1995; Joe still lives in Coffeyville, Kansas, their adopted hometown. The brothers grew up with the budding aviation industry, having been bitten by the flying bug while watching dirigibles in Akron during World War I.

In 1929, the brothers took up glider flying, a sport somewhat cheaper than powered flight. In 1931, they began to fly powered airplanes, but were still drawn to gliders because of their lower cost. Dismayed by the lack of two-seat gliders for training new pilots, the brothers decided to try their hand at building one. Their two-seater was finished in 1933, and turned out quite well, but the effects of the Depression prevented them from capitalizing on it commercially.

The Great Depression further put powered flight out of reach for the brothers, as well as for many other aspiring aviators. The brothers decided that if they could build a glider, they could build a powered airplane. In July 1934, their homemade airplane first flew, powered by a 45-hp, three-cylinder Szekely engine. The engine was so unreliable, however, that their glider experience came in handy.

After the 1932 presidential election ushered in the New Deal, the Funk brothers saw a glimmer of hope. One of the economic jump-start programs initiated by the Franklin Roosevelt administration was a competition to produce a $700 two-place airplane. The Funks decided to enter their airplane. To cut costs and to get adequate power, Howard, the mechanically gifted brother, used a highly modified Ford four-cylinder Model B car engine. Joe Funk flew the little craft to Washington, D.C., and won an order for one airplane and a spare engine.

Flushed with their prize, the Funks took the craft to the National Air

1949 Funk-Ford six-cylinder 8N
Some 5,000 Ford tractors had their regular four-cylinder engine replaced by the six- or eight-cylinder Ford truck engine using the Funk Aircraft Company's kit. The six produced 80 hp at 2,400 rpm and did not drastically alter the classic lines of the 8N. Owner: Dean Simmons.

Races in Cleveland in 1937. A group of Akron, Ohio, businessmen became interested in their plane and formed the Akron Aircraft Company to manufacture it. Joe says the selling price was not $700, but $1,700— and it should have been $2,700. Over one hundred of the Funk B (due to the Model B engine) were sold.

By 1940, Funk B sales were drying up. The Funks then switched to the new 75-hp Lycoming air-cooled aircraft engine. Sixty airplanes were sold, but not before the Akron company went into receivership. In 1941, the brothers left Akron and moved to Coffeyville, Kansas. The Jensen brothers, foundrymen from Coffeyville, picked up the receivership and production continued.

When the World War II started, the Funk's manufacturing capability was utilized to make military aircraft components. After the war, production of the Funk airplane resumed using an 85-hp Continental en-

gine. This Funk B-85-C was the final version of the airplane, and about two hundred were made between 1946 and 1948.

In 1948, Joe was traveling with one of Funk Aircraft Company's salesmen when they happened to stop at a Ford dealership in Milford, Illinois, run by Ollie E. Glover. Glover had crudely installed a Ford six-cylinder in a 9N. According to Joe, the tractor assembled by Glover used a welded adapter. "It even sagged in the middle a little," Joe remembers. He offered to make proper patterns for casting the adapter, and to design and build the other kit parts. Soon Glover was ordering large numbers of kits for which he had obtained a patent. The Funks saw that Glover was not servicing the western market, so arranged to handle sales west of the Mississippi, while Glover handled sales east of the river.

At about the same time, a pair of Iowans, Delbert Heusinkveld and Quinton Nilson, jury-rigged the first V-8 installation into a Ford tractor. Nilson entered it in the National Plow Terrace Contest at West Point, Nebraska, in 1949, and won the contest by a wide margin. After the pair had converted about a dozen tractors to V-8s, Nilson called on the Funks to see if they would make a cast-iron oil pan with an attachment for the front axle. The Funks obliged, and the rest is history.

Funk-Ford Tractor Conversions

Most of the Funk-Ford conversions used the Ford industrial version of the car and truck 226-ci (3,702-cc) six-cylinder engine that was rated at 80 hp at 2,400 rpm and produced 182 ft-lb (247 newton-meters) of torque at 1,200 rpm. A few Ford NAA tractors were converted to the new I-block overhead-valve six-cylinder that came out in the 1953 model. It produced about 100 hp as a tractor prime mover.

Less than a hundred of the kits were built using an industrial version of the standard car and truck flathead Ford 239-ci (3,915-cc) V-8. The V-8 was rated at 85 hp at 2,400 rpm for industrial use, down from the 100 hp at 3,600 rpm for automotive use. Torque of the famous V-8 was 187 ft-lb (253 newton-meters) at 1,600 rpm.

Most of the Funk conversions were done on new Ford tractors. For around $2,200, the price of a nice new car in 1952, a farmer could own the most powerful wheel-type tractor available. The big new John Deere R, Case LA, International Harvester WD-9, and the Oliver 99 all had less than 55 hp and were in the $3,000–4,000 price range. It wasn't until 1954 that production tractor power even approached 75 hp. And in that same year, a disastrous factory fire put an end to the making of Funk-Ford kits.

The Funk conversion consisted of a gearbox adapter and adapters for the front wheel attachment, which was usually a special cast-iron oil pan with an axle attachment flange, although the earliest ones used frame rails. The tractor was lengthened about 8 inches (20 cm), so tie and radius rods were lengthened, and a hood extension was provided to reach the instrument panel. The hood and grille were raised, and an appropriate radiator used. Usually, the largest available front and rear tires were used, as was all manner of ballast.

Above: **Rare 1950 Funk conversion kit brochure with handwritten changes**

1952 Funk-Ford six-cylinder 8N
Facing page, top: *The 225-ci (3,686-cc) Ford truck/industrial six-cylinder engine installed in a Funk-Ford 8N. Notice the governor mounted just behind the fan. Owner: Robert Meyer.*

1952 Funk-Ford six-cylinder 8N
Facing page, bottom: *Robert Meyer's grandfather had the first Ford-Ferguson 9N in Lake County, Illinois. Grandpa put Bob on the seat of the 9N in the field at age three. Bob couldn't reach the pedals, but knew how to kill the engine to stop. Thus, he was able to go around and around with the disc or harrow. Tractor remanufactured by N-Complete.*

Surprisingly, the Ford Motor Company did not approve of the Funk-Ford conversion, although it did not object to selling loads of industrial engines to Funk. The attitude of Ford engineers was that there was no need for such horsepower, and that the six- and eight-cylinder engines over-torqued the drivetrain. In fact, they worried that the great power increase would lead to back-flips and rollovers. At one point, Ford threatened Funk with a cease-and-desist lawsuit as a result of a rash of rear-end gear failures. Later, Ford determined that one of its factory shifts had built tractors with rear-end gears that had missed a heat-treatment step.

Did the power increase of the Funk-Ford conversion overtax the tractor's drivetrain? Well, it depends. The rear-axle mechanism was the same as that used in Ford trucks that used these same engines. The gear ratio to the rear axle was numerically higher in the tractor application, however, increasing the torque. In actual practice, Funk-Fords do not have much rear-end trouble. If the tractor is used extensively for heavy plowing or the like, gear problems can develop. Certainly, differential life will be shortened. With the original four-cylinder engines, most differential gears are still in perfect condition after more than fifty years of service, so the effect of shortened life may not be serious. With the Funk-Ford, care should be exercised when starting a dead load, or if the tractor gets stuck in heavy soil. For jobs such as mowing or others where much of the power goes through the PTO, the Funk-Ford is hard to beat and the rear end should not be overtaxed.

Since most of the Funk conversions were done to new 8Ns, one might wonder what happened to the leftover new four-cylinder engines. During World War II, Ford developed a version of its half-ton pickup to use the Ford tractor four-cylinder engine. Despite impressive gas mileage, this version never caught on in the United States. They did go on to become popular in Brazil, however, and that's where the surplus tractor engines went. Ford dealers returned the engines to Ford, which shipped them to Brazil to be used as replacements for worn-out engines.

Funk also mounted Ford tractor engines on forage harvesters, using its foundry experience to make the output adapter. Funk also did some torque converter-reverser transmissions for Massey-Harris-Ferguson.

After the fire in 1954 ended production of the tractor kits, Funk went on to market a number of successful products made in new factory buildings. In 1967, the company merged with Gardner-Denver, and this unit was later purchased by Cooper Industries. In June 1989, the Funk Manufacturing Company was purchased by Deere & Company and is now producing transmissions and castings for Big Green.

1952 Funk-Ford V-8 8N

A dramatic top view of a Funk-Ford. The twin chrome pipes on either side of the hood indicates it's a rare V-8. Best estimates suggest that of the 5,000 Funk-Fords, less than 100 were V-8s. The main reason was that the six-cylinder version had more low-rpm torque for better lugging. The V-8s were mainly aimed at PTO applications, such as powering irrigation pumps, where higher rpm could be maintained.

1950s Funk-Ford six-cylinder 8N
Above: *This Funk-Ford has an extra, ten-gallon (38-liter) gas tank to keep the six-cylinder fed with fuel. When Floyd Dominique bought this Funk-Ford six, it was a basket case.*

1952 Funk-Ford V-8 8N
Right: *The 239-ci (3,915-cc) 8BA flathead Ford V-8 engine installed in a Funk-Ford tractor. The main reason the hoods of Funk-Ford tractors are higher than standard is due to the need for a larger radiator, and Ford radiators were supplied with the engine kits. Because exhaust gasses passed through the water jacket of the flathead V-8, it required an exceptionally large radiator.*

1952 Funk-Ford V-8 8N

Above: *Here's a switch: A 1941 Ford truck with a 119.7-ci (1,961-cc) Ford tractor engine next to a 1952 Ford tractor with a Ford 239-ci (3,915-cc) truck engine. Ford built pickups and delivery vans with the tractor engine in 1941 and 1942. Portland, Indiana, tractor and implement dealer Ron Stauffer developed this Funk-like conversion kit for installing the flathead V-8 in Ford tractors. This 8BA engine came from a Ford school bus.*

1950s Funk-Ford V-8 8N

Left: *The six-cylinder engine, with its 4.00-inch (100-mm) stroke, produced 182 ft-lb of torque at 1,200 rpm. The V-8, with its 3.75-inch (93.75-mm) stroke, created 187 ft-lb at 1,600 rpm. Of the fewer than 100 Funk-Ford V-8s believed to have been built, Palmer Fossum of Northfield, Minnesota, owns two.*

Chapter 8

The Friendly Fergie

There is a baffling array of machinery, each separate contrivance performing its ingenious function, cutting labor costs and raising the vocation of the farmer from the slavish drudgery of the days of peasant serfdom to the dignity of a scientific or professional calling.
—Barton W. Currie, *The Tractor*, 1916

Above: **Ferguson TE-20 nose medallion**

1955 Ferguson TED-20
Left: *The Ferguson tractor was a rousing success—actually out-selling its Ford counterpart most years. Ferguson also developed an unbeatable line of ingenious implements. Owner: Brian Whitlock of Yeovil, Somerset, Great Britain.*

The World War I food shortage was not forgotten by the British government. It had resulted in the development of the Fordson and the importation of some 6,000 M.O.M. tractors by 1918. Twenty years later, war again loomed on the horizon. British Ford's General Manager, Patrick Hennessey, convinced the government to stock 3,000 new Fordsons at dealers around the country. If war came, the tractors would be employed in the same manner as in the first war. If war did not come, the dealers could sell the tractors to farmers and the government would be reimbursed.

Harry Ferguson was unaware of this arrangement when he made his famous Handshake Agreement with Ford in 1938. It was Ferguson's understanding that the Handshake Agreement included British production of the Ford-Ferguson 9N and a seat for himself on the board of directors of the prestigious British Ford. Once the production of the 3,000 Fordsons was underway, there was no way the Dagenham people, or the British government, could entertain the switch from the Fordson to the 9N. When he realized this some months into the war, Ferguson sent an angry letter to Henry Ford, stating his pique with British Ford, and announcing his withdrawal from the Handshake Agreement as far as British Ford was concerned. He also stated that he would manufacture tractors himself for the Eastern Hemisphere. Henry Ford's private secretary, Frank Campsell, wisely filed the letter without showing it to Ford. "We've got enough trouble around here without this," he said.

With the war's end in 1945, Ferguson arrived in London, took a suite in Claridges, London's best hotel, and bought a second-hand Rolls-Royce. Ferguson turned to Sir John Black of the Standard Motor Company in Coventry with a proposition to manufacture a tractor. Ferguson offered to provide a tractor design and tooling if Black would provide the manufacturing and the facility. Black agreed and offered an existing war plant on Banner Lane in Coventry. The problem, however, was not so much that of factory space, but of materials. Steel and iron were in such short supply in bombed-out England that they were rationed by the government only for high-priority uses.

Not one to deal with minions, Ferguson went directly to Sir Stafford Cripps, Chancellor of the Exchequer, the politician who headed the distribution of materials. Ferguson convinced Cripps that production of his tractor would bring hard currency into the country, especially American dollars. He also stressed how much food production could be increased, using much less of the critical materials than if the larger, more inefficient machines were used. Increased domestic food production would also lessen the need for food imports, which would help the country's balance of payments problem.

Cripps would not immediately commit to the materials, so to make

1955 Ferguson TEF-20

A TE-20 fitted with a diesel engine was termed a TEF-20. The Standard Motor Company–built four-cylinder indirect-injection engine was rated at 25 hp. The diesel engine in the TEF-20 was considerably different from the spark-ignition engine and is not retrofitable. Besides being longer, the starter is on the opposite side and the clutch housing is different. Owner: Brian Whitlock.

sure nothing was left to chance, Ferguson staged another of his famous demonstrations. He rented acreage near the London airport at Heathrow. In the barn on the property, Ferguson's 9N was disassembled, cleaned, adjusted, and re-assembled with the utmost care. It was then carefully repainted in the same gray as before. Some twenty dignitaries were invited, including representatives of the Chinese government. As hoped, all were favorably impressed, and all resistance to the Ferguson tractor program faded.

Debut of the Ferguson TE and TO Tractors

Ferguson's new tractor would bear a suspiciously close resemblance to the Ford-Ferguson. In the mid-1940s, Ferguson had been given a set of Ford-Ferguson 9N blueprints that were to be delivered to the industrial consultant firm of Ford, Bacon & Thomas to review manufacturing procedures to suggest possible cost cutting, as Brock remembers. Ferguson never returned the drawings, and used them and the consultant's ideas to construct his own, Ferguson tractor. He also obtained tooling from the United States, from the same sources that Ford used. In the end, Ferguson's design was a virtual copy of the Ford 9N, except that it boasted of such modern improvements as an overhead-valve engine and a four-speed transmission.

Ferguson's tractor was known as the TE-20—TE for Tractor-English, 20 for its horsepower rating. A Standard Motor Company engine was to be used, but it was not ready in time, so a four-cylinder

1940s Ferguson TE-20

When Ferguson launched his tractor production in 1946 at Banner Lane in Coventry, England, he obtained parts, patterns, and fixtures from the same sources as Ford. Thus, it's no wonder the Ferguson TE-20 looks like a 2N— right down to the paint color. This is an early, Banner Lane TE-20 Ferguson, serial number TEA 257732. Those with TEA numbers were equipped with the four-cylinder Standard engine. In 1951, power was raised from 23 to 28 hp at 2,000 rpm; because of the increase in power, these are sometimes referred to as TE-25s. Owner: Dr. Larry George of Zionsville, Indiana.

Above: **1949 Ferguson ad**

1955 Ferguson TEF-20

Left: *The TE-20 and TO-20 resembled the 2N in many ways.
Both the Continental and Standard engines initially had
virtually the same displacement as the Ford engine—120 ci
(1,966 cc) in the Ferguson and 119.7 ci (1,961 cc) in the Ford.
The Ferguson engine was an overhead valve, however, rather
than Ford's L-head. The Ferguson used a four-speed compared
to the Ford's three-speed gearbox. Weight, size, and price were
about the same. The chassis plate on this 1955 model indicates
Ferguson as the manufacturer, despite the Massey-Harris
merger of 1953.*

*For a better living and a better world through . . .
Lower production costs and increased profits for the
farmer . . . Less world unrest from hunger and
want . . . Greater security for world peace.*
—Ferguson tractor ad, 1949

120-ci (1,966-cc) engine was substituted from Continental Motors of Muskegan, Michigan. This overhead-valve engine developed 24 hp at 1,500 rpm—about the same as the 119.7-ci (1,960-cc) side-valve 2N engine. By 1947, an overhead-valve 120-ci (1,966-cc) Standard engine replaced the Continental; this was essentially the same engine as used in the Standard Vanguard automobile. With the Standard engine, the TE-20 had 28 hp available at the flywheel.

Ferguson was also in need of a tractor to sell in North America. When young Henry II ended the Handshake Agreement in 1946, it spelled disaster for Harry Ferguson, Inc., Ferguson's North American company. Ferguson now had a distribution operation without a tractor to distribute. When deliveries of Ford 2N tractors to Harry Ferguson, Inc., ended in 1947, Ferguson began importing TE-20s into North America. The success of this venture prompted him to start a production facility in Detroit to build an Americanized version, the TO-20—TO for Tractor-Overseas. The TO-20 reverted to the 120-ci (1,966-cc) Continental engine.

Ferguson's TE-20 and TO-20 were rousing successes. By 1951, production topped 100,000 units annually. Versions were built for gasoline, vaporizing oil, and in 1951, diesel fuels. Besides the regular farm tractor, versions were offered for industrial use and for use in vineyards. Ferguson's team of engineers also developed an unbeatable line of implements for use with the Ferguson System hydraulics.

In August 1951, an improved version of the Ferguson TO-20 was unveiled for the North American market. Tractor serial number 600001 switched over to a new four-cylinder overhead-valve 129-ci (2,113-cc) Continental engine that fathered 28 PTO hp at 2,000 rpm. Fitted with the larger engine, the new tractor was designated the TO-30, and it became stiff competition for the Ford 8N for years.

By 1954, the Ford NAA Jubilee tractors boasted several competitive advantages over the TO-30. Therefore, Ferguson brought out the TO-35, sporting a 134-ci (2,195-cc) version of the Continental engine. This larger engine provided 32 hp at 2,000 rpm. It also had a new gearbox with six forward speeds and a new 1,900-psi hydraulic pump with four cylinders. While its shape and size were the same as its predecessors, a new gray-and-green color scheme was used.

Ferguson Merges with Massey-Harris

Ulsterman Harry Ferguson soon tired of the tractor battles. For a decade, he had played David to Ford's Goliath. In 1953, he merged Harry Ferguson, Inc., with the venerable Canadian Massey-Harris Company of Toronto, Ontario. The new firm was renamed Massey-Harris-Ferguson, Ltd.—or to its legions of fans, simply Massey-Ferguson.

By the mid-1950s, nearly 300,000 Ford 9N and 2N tractors bore Harry Ferguson's name, followed by nearly 1,000,000 Ferguson TE and TO tractors. From Harry Ferguson's first brainstorm of integrating the tractor and implement in 1917 came the three-point hitch used on virtually every farm tractor today. Massey-Ferguson has gone on to become today's leader in the worldwide agricultural machinery business.

1955 Ferguson TED-20
Above: *The Ferguson sheet metal looks somewhat fresher and more modern than the art deco styling of the Ford 8N. Tractors must be road registered in Great Britain. Owner Brian Whitlock is at the controls.*

1955 Ferguson TED-20
Facing page, top: *The TED designation signifies the TVO (Tractor Vaporizing Oil) engine. TED-20 and TEH-20 (kerosene) tractors were built from 1949 to 1956, and are distinguished by the manifold shield. The TVO engine was rated at 23.9 hp. Owner: Brian Whitlock.*

1950s Ferguson-Selene TED-20 four-wheel drive
Facing page, bottom: *Selene of Turin, Italy, produced a small number of Ferguson tractors converted to four-wheel drive with components from Willys Jeeps. The conversion proved reliable and ideal for wet or hilly ground. Ferguson-Selene tractors were popular in Italy as well as in Scotland and Wales.*

Chapter 9

The Road to the World Tractor

Never complain, never explain.
—Henry Ford II, 1957

Above: **1958 Ford 961 ad**

1960 Ford 601 Workmaster
Left: *Ford tractors kept getting better throughout the 1950s and 1960s. Owner: Mike Hanna of Pendleton, Indiana. Tractor remanufactured by N-Complete.*

The Ford Motor Company celebrated its fiftieth anniversary in 1953, which was also a year of monumental changes at the firm. The 1953 model cars, trucks, and tractors were all significantly updated; these were the first models planned, styled, and engineered by Henry II's new management team. In May 1953, the Ford Motor Company acquired all the assets of Dearborn Motors, Inc., and made it a division of the parent firm.

The 1953 Model NAA tractor was Ford's first completely redesigned tractor since the 9N of 1939. Placed above the grill on the NAA was a prominent circular emblem with the words "Golden Jubilee Model 1903–1953." "Golden" referred to the company's fiftieth anniversary; "Jubilee" is a Biblical term relating to fifty-year periods. The NAA has since been known simply as the Jubilee model.

Ford engineers had already planned an improved tractor for the company's fiftieth anniversary by the time the Ford-Ferguson lawsuit was settled in 1952. The new Jubilee began production in January 1953.

From front to back, the Jubilee differed dramatically from the N Series tractors. The NAA was 4 inches (100 mm) longer, 4 inches (100 mm) higher, and 100 lb (45 kg) heavier than the Ford 8N that it replaced. It sported an all-new four-cylinder overhead-valve engine. It was completely restyled to no longer resemble previous Ford tractors—nor could the NAA be mistaken for the 1951 Ferguson TO-30, which was a competitor of the 8N.

The Jubilee's new engine was obviously targeted at the TO-30. Ford's Red Tiger engine was an inline four-cylinder displacing 134 ci (2,195 cc) from a 3.44x3.60-inch (86x90-mm) bore and stroke. The overhead-valve engine produced a maximum of 30.15 hp on the belt at 2,000 rpm during University of Nebraska Tractor Tests; this made it comparable to the Ferguson's 134-ci (2,195-cc) Continental engine, which provided 32 hp at 2,000 rpm. The Jubilee was nominally considered a two-plow tractor capable of pulling two 16-inch (40-cm) bottoms, but three 14-inch (35-cm) bottoms were commonly applied.

Other needed improvements over the 8N were also added to the Jubilee. The NAA had a better governor, new rear-axle seals and brakes, and a temperature gauge included as standard equipment on the instrument panel. The muffler was relocated alongside the engine to a spot under the hood; this allowed for an optional vertical exhaust and reduced the possibility of the muffler causing a fire in dry straw. Jubilee tractors were painted the same as the 8N.

To get around Ferguson's hydraulic-control patents, the NAA incorporated a new engine-driven hydraulic pump with "downstream" regulation. The Jubilee first used a vane pump mounted under the hood, alongside the engine. The pump had a flow control, called the Hy-trol. A

1953 Ford NAA Golden Jubilee brochure

piston-type pump was later substituted. A separate hydraulic reservoir was provided, as were provisions for remote hydraulic cylinders.

The Jubilee included a non-live PTO as standard, but an unusual live PTO was optional. A hydraulic clutch operated the live PTO. The clutch was operated by a separate pump, which was operated by the cable that drove the ProofMeter, which was a tachometer and hours-worked gauge. Needless to say, this was not a common option.

The 1954 version of the NAA was still called the Jubilee. Its nose medallion merely had stars encircling the emblem, not the "Golden Jubilee" wording. Internally, the 1954 version had gear-ratio changes that produced a reduction in operating speed for a given engine rpm.

Proliferation of the Ford Tractor Line

In September 1954, one of the most significant events in the life of the Ford Motor Company occurred: In order to raise capital, Henry Ford II announced that Ford was going public. For the first time since 1919, Ford would soon be selling its stock to the public on the open market. The impact of the stock sale, which took place early in 1955, was that outsiders now had a voice in what went on at Ford. Henry II now had to answer to the stockholders. There could no longer be altruistic motives in product planning; every division had to pull its own weight. The first casualty was the Continental Division headed by Henry II's brother, Benson. The luxurious Continental Mark II was intended as a loss-leader prestige car, but such things were hard to explain to stockholders, so the division was folded into Lincoln, and the "American Rolls-Royce" faded.

1953 Ford NAA Golden Jubilee
Ford Motor Company celebrated its fiftieth anniversary in 1953 with new car, truck, and tractor models. The new Ford NAA tractor had a round medallion on its nose incorporating a new tractor logo: a corn stalk in the center of a shield topped by the Ford script in an oval, and the words, "Golden Jubilee Model 1903–1953." The 1953 Jubilee can be distinguished from the 1954 and later Model 600s by the nose medallion. Only the '53s had the Golden Jubilee words. Both the 1953 and 1954 versions are Model NAA tractors and both are commonly called Jubilees. Owner: Floyd Dominique.

1953 Ford NAA Golden Jubilee

Above: *The NAA Jubilee featured more-complete and better-displayed instrumentation. On top is the tachometer/hour meter. Below are the ammeter, coolant temperature, and oil pressure gauges.*

1953 Ford NAA Golden Jubilee

Left: *The NAA was the first all-new tractor by Ford in fourteen years. It was larger, heavier, and more powerful than the 8N it replaced. Changes included new front wheels, slower reverse gear, stronger front axle, larger fuel tank, and a hidden radiator cap under a full-length hood hatch. In 1953, the new Ford NAA sold for $1,745. Owner: Don Sparks of Shirley, Indiana. Sparks's father bought the tractor new.*

1955 Ford 600

Above: *In 1955, Ford for the first time offered a second tractor line, and this 600 was one of 66,656 U.S.-built Ford tractors that year. The two-to-three-plow utility tractor was little changed from the NAA Jubilee of 1954. This style of grille was used from 1955 to 1961 on the 600 Series Fords. Tractor remanufactured by N-Complete.*

Right: **1955 Ford 600 and 800 brochure**

The Tractor and Implement Division was also difficult to justify, since independent Ford tractor operations were at work in North America and the United Kingdom. To make matters worse, in 1954, British production came within 5,000 units of U. S. production. In 1956, Dagenham out-produced Dearborn.

The response of the Whiz Kids management team was to broaden Ford's tractor line. For forty years, Ford had followed a policy of building just one tractor model, both in North America and England—albeit separate models for the two markets. Times were changing. Farms in North America and abroad were growing larger. Ford tractor dealers did not have the products to support themselves in the changing market.

Thus for 1955, two new tractors were offered by Ford. The first was the Model 600, with the same Red Tiger 134-ci (2,195-cc) engine as the Jubilee. The 600 was offered in three variations: the 640 featured the NAA's four-speed transmission and non-live PTO; the 650 had a new five-speed and the non-live PTO; and the 660 had the five-speed and a live PTO.

The second tractor, the Model 800, used the same four-cylinder engine, but it was bored out to 172 ci (2,817 cc). The more powerful Model 800, which was Ford's first American-built three-plow tractor, was available as the 850 and 860, both with the new five-speed gearbox. The 850 had a non-live PTO, the 860 had a live PTO. All of the 600 and 800 versions were of the utility-tractor configuration and were equipped with live hydraulics and a three-point hitch. The models proliferated in 1956 with three Special Utility versions using the old four-speed gearbox: the 620 and 820 without three-point lift or PTO, and the 630 version with lift but no PTO.

In 1956, new 700 and 900 series tractors were added, which were the same as the 600 and 800 except they were of tricycle configuration. These new series included the 740 with the 134-ci (2,195-cc) engine, and the 950 and 960 with the 172-ci (2,817-cc) engine. This line-up was continued through 1957.

Ford tractors received a styling facelift and improved features before the end of the 1957 model year. To signify the improvements, a "1" was added to each model designator, changing the 960 into the 961, for example.

To further differentiate its lines, Ford labeled tractors with the 134-ci (2,195-cc) engine as the Workmaster series; tractors with the 172-ci (2,817-cc) engine were the Powermaster series. Diesel Workmasters and

1955 Ford 800

The 800 Series complemented the 600 Series, which was a continuation of the NAA. The 800 was available as the 860 with a live PTO or as the 850 without. The Ford 861 was built from 1958 through 1961. It used the 172-ci (2,817-cc) four-cylinder engine available in LPG and diesel versions, as well as the gas type shown here. It was rated a three-to-four-plow tractor. Owner: Floyd Dominique.

1961 Ford 971 LPG Workmaster

Above: *The Workmaster tractors with the 172-ci (2,817-cc) engine reverted to the 8N red-and-gray scheme in 1959. Also in 1959, the Select-O-Speed power shift transmission was introduced. Ford introduced LPG tractors in 1957. The fuel was popular for a time because it allowed higher compression ratios and thus, more power from a given displacement. The clean-burning fuel also contributed to much longer engine life. Owner: Dwight Emstrom.*

1959 Ford 961 Diesel

Right: *The Ford 900 Series, introduced in 1955, bore the model designation NDB. The 961 Powermaster was Ford's first row-crop tractor since the Fordson All-Around of 1937. Prior to 1959, only gas and LPG engines were available in the five-speed 961. Owner Floyd Dominique bought this 961 new.*

Powermasters based on the existing Red Tiger engines became available in 1958 and 1959, respectively. Liquefied petroleum gas (LPG) variations of each engine were also offered.

In 1959, several other changes were made to the lineup. Select-O-Speed power-shift planetary transmission was first offered, providing ten forward and two reverse speeds. The Workmaster's color scheme was switched to red with gray trim. The Powermaster line retained the gray sheet metal with red cast iron, but added a red center strip on the hood and red grill. Also in 1959, high-crop off-set cultivating tractors were offered as the 501 Workmaster series.

Ford had indeed met its goal of breaking away from being a one-tractor company. Still, Ford lost market share throughout the late 1950s. In 1956, U.S. tractor production actually dropped below that of British Ford; this trend continued until the end of Fordson production in 1964, and in some years, English production was triple that of North American output.

The World Tractor

Possibly to hide Dearborn's failure to out-produce Dagenham, the two operations were merged into the newly established Ford Tractor Operations in March 1961. This was the first step in a plan to make a single line of tractors for worldwide distribution. Internally at Ford, the concept was known as the "World Tractor."

Concurrently, a new six-cylinder Model 6000 tractor was introduced. In design over the previous three years, the 6000 was unlike anything built before by either Dagenham or Dearborn. It was designed to be a row-crop tractor, although it was available with either a wide-front or tricycle-front wheels. Its engine was available in gasoline, diesel, or LPG versions. With a working weight of 10,000 lb (4,500 kg), the 6000 was large and needed all of its 60 hp on the drawbar horsepower. Although it continued in production until 1967, the 6000 was never popular, and it was sold only in the United States, Canada, and Australia.

Ford Tractor Operations was renamed the Ford Tractor Division in early 1962. The numbers of the agricultural tractors were all changed that year to the four-digit scheme. The Fordson Super Major was given the Ford 5000 designation and the Super Dexta was called the Ford 2000 Diesel. The paint combinations were also changed so that all tractors were painted in variations of blue and gray, rather than the North American tractors' red and gray and the British blue and orange. These were stop-gap measures to carry them over until the World Tractors could be readied.

In 1964, tractor production ended at Dagenham after thirty-one years, and with it, forty-seven years of the Fordson name. Tractor production in Dagenham's last year was more than double that of American-built

The farmer must either take up power or go out of business.
—Henry Ford, *My Life and Work*, 1926

1957 Ford 950-4
The 950-4 designation signifies a high-clearance row-crop tractor with adjustable width on all four wheels and with a non-live PTO. Owner: Dwight Emstrom.

145

1965 Ford-Northrop 5004

Above: *The Chaseside Company, a loom manufacturer in Blackburn, Lancashire, Great Britain, built four-wheel-drive Northrop conversions of Ford tractors between 1965 and 1967. These conversions were built on the Ford "skid unit" and based on the 5000 tractor. Northrop production would have continued if the British JCB Company had not bought out Chaseside in 1968 and ceased production. Northrop engineer David J. B. Brown joined the Muir-Hill firm and developed a similar MH101 tractor. Muir-Hill is today owned by Lloyd Loaders of Hipperholme and continues to build its Myth-Holm four-wheel-drive tractor based on the New Holland 40 Series. Owner: Duke Potter.*

1964 Ford 4000 Select-O-Speed

Left: *The Ford Tractor Division began consolidating its worldwide product line in 1962 by announcing its Long Blue Line. The 4000 was much the same as the 801 under the revised sheet metal and new color scheme. Owners: Ron and Shirley Stauffer.*

1996 New Holland 8340

Above: *A state-of-the-art New Holland 8340 tows a Parmiter baler through an English field with the famous Glastonbury tower in the background. The 8340 PowerStar is rated at 112 PTO hp and has a wide range of applications on mid-sized beef and dairy farms.*

1972 Ford 7000

Right: *The 1972 Ford all-purpose 7000 Diesel was equipped with a 256-ci (4,193-cc) four-cylinder turbocharged engine. It produced 85 hp at 2,100 rpm. A manual eight-speed transmission was standard. Normal operating weight was about 10,000 lb (4,500 kg). Owner: Frank Bissen.*

1979 Ford FW30

Ford's FW-30 was a big tractor for its time. Working weight approached 32,000 lb (14,400 kg), and maximum power was in the 205-hp class. A 903-ci (14,791-cc) Cummins V-8 diesel provided the power. Owner Dale Bissen of Adams, Minnesota, added a turbocharger to this tractor, making it the equivalent of the FW-60 model with 270 hp. A twenty-speed transmission was provided. The FW Series tractor was built for Ford by Steiger.

tractors, which were then assembled at the Highland Park plant. A new tractor plant was opened in Basildon, England. The Tractor Division converted the Antwerp, Belgium, plant to tractor production at the same time.

The new World Tractor line debuted in October 1964, announced for the 1965 model year. Although the number identifiers were retained, completely new tractors replaced the previous 2000 through 5000 series. Except for the row-crop 6000, row-crop and offset tractors were eliminated from the line. Innovative new three-cylinder engines replaced the four-cylinder units in the 2000, 3000, and 4000 series. The Model 5000 got a four-cylinder derivation of the new engine, while the 6000 was essentially unchanged except for its new designation as the Commander 6000.

The new tractors were much more like the British Dexta and Super Major than they were like the North American Jubilee derivatives. In fact, outside the United States and Canada, the 2000 was known as the Dexta 2000, and the 5000 as the Super Major 5000. Industrial versions had "500" added to the model number, such as the 4500 Industrial.

In 1966, the Ford Tractor Division began marketing a line of lawn and garden equipment manufactured by Jacobsen of Racine, Wisconsin. In subsequent years this segment of the tractor market was to become key to Ford.

Onward to the Millennium

By 1966, Ford Tractor Division was number two in sales worldwide. The number-one company was Massey-Ferguson, the company derived from the merger of Massey-Harris and Ferguson. After the merger, Harry Ferguson retired from the tractor business, and died in 1960. He would have enjoyed seeing Ford play catch-up.

The Ford Tractor Division celebrated sixty years of tractor production in 1977. The occasion was marked by entry into the monster four-wheel-drive articulated-tractor market when Ford contracted with the Steiger Tractor Company of Fargo, North Dakota.

In 1986, Ford purchased the Sperry New Holland Company of New Holland, Pennsylvania. The new Ford New Holland replaced the Ford Tractor Division and was headquartered in New Holland. In 1987, Ford purchased the Canadian Versatile Tractor Company of Winnipeg, Manitoba, and folded it into Ford New Holland.

In 1991, the New Holland Holding Company was formed. The assets of Ford New Holland and Fiat Agri of Turin, Italy, were placed under its control. Ford owned 20 percent of the holding company, and Fiat 80 percent. In 1992, Fiat, through a $600 million infusion of cash, increased its share to 88 percent. In 1993, Fiat completed the buyout from Ford.

Today, the tractors built by New Holland work on farms around the globe. New Holland tractors boast a wide range of applications, state-of-the-art features and a level of quality that will lead the field into the next century.

1996 New Holland 7840

Above: *One of the new generation of New Holland tractors, this 1996 7840 works with a 1986 New Holland combine on the Biss family farm in Kingweston, Somerset, Great Britain. The 90-PTO-hp Model 7840 PowerStar was built from 1991 to 1996.*

1996 New Holland 8970

Left: *New Holland's latest generation of farm tractors will carry on the Ford tradition into the next century. The 8970 Genesis tractor established the transition between the Ford and New Holland brands.* (Photo © New Holland)

Appendices

Tractor Specifications

Model	Year	Bore and stroke (inches/mm)	Cylinders	Displacement (ci/cc)	Rated Rpm	Forward Speeds	Basic weight (lb/kg)	Fuel
Fordson	1918–1927	4.00x5.00/100x125	4	251/4,111	1000	3	2,700/1,215	kerosene
Fordson	1933–1946	4.12x5.00/103x125	4	267/4,373	1100	3	3,600/1,620	kerosene
E27N Major	1946–1953	4.12x5.00/103x125	4	267/4,373	1200	3	4,000/1,800	kerosene
9N	1939–1942	3.19x3.75/80x94	4	119.7/1,961	2000	3	2,340/1,053	gasoline
2N	1942–1946	3.19x3.75/80x94	4	119.7/1,961	2000	3	2,340/1,053	gasoline
8N	1947–1952	3.19x3.75/80x94	4	119.7/1,961	2000	4	2,410/1,085	gasoline
Major	1953–1958	3.74x4.52/93.5x113	4	199/3,260	1600	6	5,100/2,295	gas, LPG and/or diesel
Power Major	1958–1960	3.94x4.52/98.5x113	4	220/3,604	1600	6	5,300/2,385	diesel
Super Major	1961–1964	3.94x4.52/98.5x113	4	220/3,604	1700	6	5,300/2,385	diesel
Dexta	1958–1961	3.50x5.00/87.5x125	3	144/2,359	2000	6	3,000/1,350	diesel
Super Dexta	1962–1964	3.60x5.00/90x125	3	153/2,506	2250	6	3,000/1,350	diesel
NAA Jubilee	1963–1964	3.44x3.60/86x90	4	134/2,195	2000	4	2,510/1,130	gasoline
600	1959–1961	3.56x3.60/89x90	4	144/2,359	2000	4/5/10	3,375/1,519	diesel
800	1954–1961	3.90x3.60/97.5x90	4	172/2,817	2000	4/5/10	3,450/1,553	gas, LPG and/or diesel
2000	1965–1968	4.20x3.80/105x95	3	158/2,588	2000	4/8	4,000/1,800	gas, LPG and/or diesel
3000	1965–1968	4.20x4.20/105x105	3	175/2,867	2000	8/10	4,100/1,845	gas, LPG and/or diesel
4000	1965–1968	4.40x4.20/110x105	3	192/3,145	2200	8/10	4,770/2,147	gasoline
5000	1965–1968	4.23x4.20/106x105	4	233/3,817	2100	8/10	5,830/2,624	gas, LPG and/or diesel
6000	1961–1967	3.62x3.60/90.5x90	6	223/3,653	2300	10	7,000/3,150	gasoline
6000	1961–1967	3.62x3.90/90.5x90	6	242/3,964	2230	10	7,165/3,224	diesel

Nebraska Tractor Tests Summary

Model	Year	Test Number	Fuel	Max Hp Belt/PTO	Max Hp Drawbar	Max Pull (lb/kg)	Fuel Consumption	Weight (lb/kg)	Wheels
Fordson	1920	18	kerosene	18.2	9.34	2,187/984	7.32	2,710/1,220	steel
Fordson	1926	124	kerosene	22.3	12.3		9.63	3,175/1,429	steel
Fordson N	1930	173	kerosene	23.2	13.6	3,289/1,480	7.07	3,820/1,719	steel
Fordson N	1930	174	gasoline	29.1	15.5		9.53	3,800/1,710	steel
9N	1940	339	gasoline	23.1	12.8	2,236/1,006	9.74	3,677/1,655	rubber
8N	1950	443	gasoline	25.5	20.8	2,810/1,265	11.2	4,043/1,819	rubber
NAA Jubilee	1953	494	gasoline	32.4	26.8	3,232/1,454	11.2	4,389/1,975	rubber
Major	1953	500	diesel	38.5	34.2	5,315/2,392	15.5	7,890/3,550	rubber
660	1955	561	gasoline	35.2	29.8	3,859/1,736	11.3	4,917/2,213	rubber
851	1958	640	gasoline	50.2	43.28	5,033/2,265	11.7	6,855/3,085	rubber
851	1958	654	diesel	44.5	38.9	5,120/2,304	14.4	6,885/3,098	rubber
641	1959	686	diesel	31.8	28.5	4,230/1,904	15.0	5,897/2,654	rubber
6000	1961	784	gasoline	66.9	59.4		10.4	9,535/4,291	rubber
Super Dexta	1963	844	diesel	38.8	32.3		15.5	6,030/2,714	rubber
3000	1965	881	diesel	39.2	34.9		16.8	6,885/3,098	rubber
5000	1966	932	gasoline	60.4	50.9		11.9	9,650/4,343	rubber
2000	1967	959	gasoline	31.2	27.5		16.0	5,860/2,637	rubber

Notes:

Belt/PTO Hp: This is Test C horsepower, maximum attainable at the PTO or belt pulley. If the generator, hydraulic pump, etc., were not standard equipment, they were removed for these tests.

Drawbar Hp: Taken from Test G data, it is based on maximum drawbar pull and speed. The difference between this and PTO Hp is due to slippage, and to the power required to move the tractor itself. The heavier the tractor, the less the slippage, but the more power required to move the tractor. Factory engineers looked for the ideal compromise.

Max. Pull: Test G.

Fuel Consumption: The rate of fuel consumption in horsepower hours per gallon.

Weight: The weight of the tractor plus ballast in pounds and kilograms. Ballast was often added for Test G and other heavy pulling tests, and then removed for other tests to improve performance.

Serial Numbers

The surest way to tell the year that a Ford tractor was made is by the serial number. On the Fordson, you will find the number on the right side of the engine, between the front two ports of the manifold, just below the cylinder head. This applies to all Fordsons through the E27N model, except for those with Perkins diesel engines. Note that these numbers were hand stamped and may be uneven and not clearly imprinted.

Don't forget that the first of these tractors were the fifteen X models. These were numbered X-1 through X-15 and were built in 1917. Later in 1917, production began for the British Ministry of Munitions. These M.O.M. tractors began with serial number 1 and ran through 259 by the end of 1917. During 1918, M.O.M. deliveries ran through serial number 3900.

The M.O.M. tractors were not yet Fordsons. The name "Fordson" began to be used about April 1918. These first Fordsons were known as Model F Fordsons. Except for the first ten built for Henry Ford's special friends (numbered 1 through 10), Fordsons kept on with the numbering established for the M.O.M. order. You can cross-check the engine numbers of M.O.M. tractors and Models F and N against some distinguishing characteristics, as noted below.

Ford tractors have their serial numbers on the engine block on the right side

Fordson Model F

Year	U.S. Production	Irish Production
1917	1 to 259	
1918	260 to 29979	
1919	34427 to 88088	63001 to 63200
		65001 to 65103
1920	100001 to 158178	65104 to 65500
		105001 to 108229
1921	158312 to 170891	108230 to 109672
1922	201026 to 262825	109673 to 110000
		170958 to 172000
		250001 to 250300
		253001 to 253552
1923	268583 to 365191	
1924	370351 to 448201	
1925	455360 to 549901	
1926	557608 to 629030	
1927–1928	629830 to 747681	

Fordson Model N

Year	Cork Production Beginning Numbers
1929	747682
1930	757369
1931	772565
1932	776066

Year	Dagenham Production Beginning Numbers
1933	779154
1934	781967
1935	785548
1936	794703
1937	807581
1938	826779
1939	837826
1940	854238
1941	874914
1942	897624
1943	925274
1944	957574
1945	975419

Fordson E27N Major

Year	Beginning Serial Number
1945	80520
1946	93489
1947	1018979
1948	1054094
1949	1104657
1950	1138033
1951	1180610
1952	1216575

Records indicate the last E27N was serial number 1216990

New Major

Year	Beginning Serial Number
1953	1247381
1954	1276857
1955	1322525
1956	1371418
1957	1412409
1958	1458381

Power Major

Year	Beginning Serial Number
1958	1481091
1959	1494448
1960	1538065
1961	1583906

Super Major

Year	Beginning Serial Number
1961	08A 300001M
1962	08B 740000A
1963	08C 781370A
1964	08D 900000

Fordson Dexta

Year	Beginning Serial Number
1958	16066
1959	20427
1960	46212
1961	72003

Fordson Super Dexta

Year	Beginning Serial Number
1961	09A 312001M
1962	09B 070000A
1963	09C 731454A
1964	09D 9000000A

Ford-Ferguson

Model 9N

Year	Beginning Serial Number
1939	1
1940	10234
1941	45976
1942	88888

Model 2N

Year	Beginning Serial Number
1942	99003
1943	105375
1944	126538
1945	169982
1946	198731
1947	258504

The last serial number 2N is thought to be 296131

Model 8N

Year	Beginning Serial Number
1947	1
1948	37901
1949	141370
1950	245637
1951	343593
1952	442035

The last 8N serial number is thought to be 524076

Model NAA

Year	Beginning Serial Number
1953	NAA 1
1954	NAA 77475

The last serial number NAA is thought to be NAA128965

600, 700, 800, 900 Series

Year	Beginning Serial Number
1954	1
1955	10615
1956	77615
1957	116368

501, 601, 701, 801, 901 Series

Year	Beginning Serial Number
1957	1001
1958	11977
1959	58312
1960	105943
1961	155531

2000 and 4000 Series

Year	Beginning Serial Number
1962	1001
1963	11949
1964	38931

6000 Series

Year	Beginning Serial Number
1961	131591
1962	171542
1962 (Blue Line)	1001
1963	11948
1964	38931

2000, 3000, and 4000 Series

Year	Beginning Serial Number
1965	C100000
1966	C123000
1967	C160000
1968	C188000

Commander 6000 Series

Year	Beginning Serial Number
1965	C100000
1966	C123000
1967	C160000

Model Characteristics

M.O.M. tractors 1917–1918
Shallow-lid toolboxes with no logo
Cast-iron front wheel hubcaps
Tie rods and steering rods were three pieces
Six-spoke rear wheels
Gas tank had no logo and was rounder in shape
No top center seam on tank
No logo on the radiator tank
Ladder-side radiator shell
Oil filler in rear of engine, held on with two bolts

Fordsons 1918–1919
Six-spoke rear wheels
No fenders or fender mounting provisions
Ladder-side radiator cover with "Fordson" cast on the front
"Manufactured by Henry Ford & Son . . ." on the rear of the fuel tank
Model T coil box with cast brackets
Round axle housings without grooves
Maple wood steering wheel as on Model T
Pressed-steel seat with slotted holes
Three-hole drawbar

Fordsons 1920–1923
Seven-spoke rear wheels
Solid-side radiator cover with "Fordson" cast on the front
"Manufactured by Ford Motor Company . . ." on the rear of the fuel tank
"Rainbow" coil box
Round axle housings without grooves
No fenders or fender mounting provisions
Maple wood steering wheel
Pressed-steel seat with round holes
Five-hole drawbar
Two oil drain plugs (one for dirt trap)
Toolbox with large "Fordson" logo and extended flanges
1922 and on had a double-lead worm final drive
1923 and on had a transmission disk brake

Fordsons 1924–1926
Double-lead worm final drive
Fenders optional
Toolbox fenders available in 1925
Solid-side radiator cover with "Fordson" logo cast on the front
New "Fordson" coil box in 1925
Hard rubber steering wheel
Five-hole drawbar
Two-bung fuel tank (gas starting tank internal, eliminating the cast fuel tank bolted to the water washer)
Model T–type oil breather
Oil filler had flip-up cover
High-bead front wheels

Fordsons 1927
Unchanged except for a revised engine block with more water cooling for the valves
Lower manifold assembly

Cork Model N Fordsons 1929–1932
Engine bore increased
Conventional impulse magneto added
Water pump added
Oil-bath air cleaner added
Heavy front axle with downward bend in the middle
Cast-iron front wheels
Five rounded triangular holes in the front wheel disks
"Made in the Irish Free State" or ". . . in Cork, Ireland" on fuel tank end
Tapered toolbox fenders standard equipment

Dagenham Fordsons Model N 1933–1945
"Made in England" on fuel tank end
Plain fenders standard equipment, tool box on dash
Pneumatic tires optional after 1935
All-blue paint until 1935
Blue with orange trim from 1935 to 1937
All-orange paint from 1937 to 1939
All-green paint from 1939 to 1945
All-Around row crop version available in 1937

9N Ford-Ferguson Tractor 1939
Aluminum hood and side panels on first 700–800 tractors
Key and starter button on the dash
An "Ignition-On" light on the dash
Grease fittings on the front of the kingpin housings
Two-ribbed fenders, bolted/riveted to brackets
Four-spoked steering wheel
Smooth rear-axle hubs
Front-axle radius rods are I-beam type
Four-bladed pusher fan
No freeze plugs on side of engine block
Square exhaust manifold cross-section
Battery/fuel tank filler cover is a snap-in (not hinged)
Cast-aluminum grille with semi-horizontal spokes
Left and right brake pedals are interchangeable
Extensive use of aluminum castings for dash and steering housing, battery stand, lift quadrant, transmission cover and inspection plates

9N Ford-Ferguson Tractor 1940
Safety Starter introduced mid-year
Ignition key on right side of dash, later moved to left side
Hinged battery cover replaces the snap-in mid-year
Three-brush generator standard
Fenders with single ribs introduced mid-year
Freeze plugs in sides of engine block
Some steel castings replace aluminum
Exhaust manifold more rounded in cross-section

9N Ford-Ferguson Tractor 1941
Left and right brake pedals now different
Kingpin grease fittings moved to the rear
Steel grille with vertical bars and solid center
Three-spoke steering wheel introduced mid-year with covered, or solid, spokes

Liberal use of chrome trim
Hubs now riveted rather than smooth
Heavier-duty lift cylinder and spring
Ignition key moved to the steering column
Six-blade fan replaces four-blade (can be pusher or puller)
Aluminum dash castings now steel
Lube provided to governor by line from oil filter

9N Ford-Ferguson Tractor 1942
1942 9Ns are similar to the 1941 model. As parts were used up, 2N characteristics appeared

2N Ford-Ferguson 1942
Slotted grille center bar
Three-spoke steering wheel with rod spokes
10.00x28-inch (25x70-cm) tires standard
8.00x32-inch (20x80-cm) tires optional
Valve rotators used
Non-electrical, steel-wheeled version produced
External fasteners in lower side panels mid-year
Chrome trim replaced by black paint
Fender bolt holes in rear axle no longer solid as of mid-year
"2N" appears on lower edge of oval Ford badge

2N Ford-Ferguson 1943
Radiator pressurized at serial number 109502
9.00x32-inch (22.5x80-cm) tires optional
8.00x32-inch (20x80-cm) tires not offered

2N Ford-Ferguson 1944
Oval radius rods introduced mid-year
Sealed-beam headlights available
Helical transmission gears introduced mid-year, HX marked on case
Identifying year number on flywheel housing

2N Ford-Ferguson 1945
Heavier rear-axle housings introduced mid-year
Rear-axle castings dated

2N Ford-Ferguson 1946
Still heavier rear-axle housings used

2N Ford-Ferguson 1947
No changes

8N Ford July 1947
Differences between 8N and previous 2N tractors:
Compression ratio, 6.0:1 early (same as 2N), 6.7:1 late (gasoline), 4.75:1 for 8NAN (kerosene)

Independent brakes on each rear wheel controlled by pedals on the right side of the tractor; no means to lock pedals together, although both pedals could be depressed together by the right foot. No clutch pedal interconnect to brake.
Besides draft control, a toggle lever under the seat can select "Position Control." With this lever actuated, draft control is blocked out and implement position is controlled strictly by the quadrant
Light gray sheet metal and bright red cast iron.
Recirculating-ball steering replacing the sector gears of the Ford-Ferguson
Higher steering wheel, flip-up seat, and running boards allowed for operation while standing.
Rear wheels with rounded center disc, rather than the flat type used on the Ford-Ferguson; same type rim
Front and rear wheels now bolted at hub, rather than being bolted to a plate hub as on the Ford-Ferguson
Screened-grille engine air intake positioned on the right aft hood
Ford script trademark embossed on the front of both sides of the hood
Ford oval emblem in the front center of the hood was now larger than that of the Ford-Ferguson and was chrome with a red background
Tip-out grille for radiator cleaning

8N Ford 1948
Different clutch linkage used
Top-link rocker, which actuates the draft control, now has three-moment arm positions rather than one

8N Ford 1949
Adjustable recirculating-ball mechanism after serial number 216998

8N Ford 1950
6.00x16-inch (15x40-cm) front tires optional
Chrome front hood emblem replaced by argent
Side distributor/coil after serial number 263844
Removable shift knob replaces knob cast with lever
Over/underdrive auxiliary transmission optional
ProofMeter tachometer at serial number 290271

8N Ford 1951
Ford script trademark embossed on fenders as well as hood

8N Ford 1952
Improved rear-axle seal introduced mid-year
Rear axle has a bulge at lower outer extremity
High-Direct-Low auxiliary transmission became a Ford-supplied option.

Recommended Reading

Arnold, Dave. *Vintage John Deere*. Stillwater, MN: Voyageur Press, 1995.

Booth, Colin E., and Allan T. Condie. *The New Ferguson Album*. Carlton, Nuneaton, Great Britain: Allan T. Condie Publications, 1986.

King, Alan C., ed. *The Fordson and Ford Tractor*. N.p., 1989.

Morland, Andrew. *Ford & Fordson Tractors*. Osceola, WI: Motorbooks International, 1995.

Pripps, Robert N., and Andrew Morland. *Ford Tractors: N Series, Fordson, Ford and Ferguson, 1914–1954*. Osceola, WI: Motorbooks International, 1990.

Pripps, Robert N., and Andrew Morland. *Fordson Tractors*. Osceola, WI: Motorbooks International, 1995.

Pripps, Robert N. *Illustrated Ford & Fordson Tractor Buyer's Guide*. Osceola, WI: Motorbooks International, 1994.

Sanders, Ralph W. *Vintage Farm Tractors*. Stillwater, MN: Town Square Books/Voyageur Press, 1996.

Vintage Tractor Special. *American Fordson & Ford 1917–1970*. Carlton, Nuneaton, Great Britain: Allan T. Condie Publications, 1991.

Vintage Tractor Special. *Ford Tractors 1964–75*. Carlton, Nuneaton, Great Britain: Allan T. Condie Publications, 1994.

Vintage Tractor Special. *Fordson Major Model 'E27N' 1945–52*. Carlton, Nuneaton, Great Britain: Allan T. Condie Publications, 1991.

Organizations

Clubs and Newsletters

9N–2N–8N Newsletter
Box 235
Chelsea, VT 05038

Antique Power
P.O. Box 838
Yellow Springs, OH 45387

Ferguson Journal
Denehurst, Rosehill Road
Stokeheath, Market Drayton
TF9 2JU Great Britain

Ford/Fordson Collectors Association
645 Loveland-Miamiville Road
Loveland, OH 45140

Gas Engines
Ironmen Album
Stemgas Publishing Company
P.O. Box 328
Lancaster, PA 17603

Wild Harvest (Massey-Ferguson)
1010 South Powell
Box 529
Denver, IA 50622

Restoration Shops and Parts Suppliers

Emstrom Farm Antiques
RR2, Box 140
Galesburg, IL 61401

Ford Tractor Specialty
Marlo Remme
RR1, Box 281
Dennison, MN 55018

K&K Antique Tractors
RR3, Box 384X
Shelbyville, IN 46176

N-Complete
10750 East 550 North
Wilkinson, IN 46186

Palmer Fossum Fords
10201 East 100th Street
Northfield, MN 55057

Strojney Implement
Mosinee, WI 54455

Wengers, Inc.
251 South Race Street
Meyerstown, PA 17067

Index

About the Author

Robert N. Pripps on an unrestored N Series Ford tractor.

Robert N. Pripps was born in 1932 on a small farm in northern Wisconsin. Besides farming, his father did local road building and maintenance with a JT crawler and Russell grader. Bob's first driving experience came at age nine when he disked a fire lane with an Allis-Chalmers crawler. During summers throughout World War II, Bob worked on neighboring farms, earning the opportunity to drive one farmer's new Farmall H at age eleven.

Bob's curiosity with things mechanical almost cost him his life at age twelve. He got a mitten caught in the power takeoff of a Gallion road grader. He extricated himself from the machine before it killed him, but the encounter cost him his right thumb.

At age fourteen came another life-changing event: His best friend's father bought the first Ford-Ferguson tractor in the area. Abject envy is not a pretty thing, but that's what reigned in Bob's heart.

Bob went to high school in Eagle River, Wisconsin, earning his private pilot's license by the time he graduated in 1950. He attended Parks Air College to study engineering and receive a commercial pilot's license and multi-engine rating. Bob took a night job at McDonnel Aircraft in St. Louis, but marriage and family responsibilities soon made the job a priority and schooling secondary. Bob became a flight test engineer on the RF-101 Voodoo while continuing night and correspondence school.

After seventeen years of part-time classes, Bob graduated from college in 1969 with a Bachelor of Science in Marketing. Bob also held a certificate in Aeronautical Engineering by that time. He then served as the marketing manager for Sundstrand Corporation's Dayton, Ohio, office.

Along the way, Bob inherited thirty acres of maple forest that were part of the Wisconsin farm on which he was born. That's when he found justification for the Ford-Ferguson 2N that helps with harvesting sap for maple syrup. He later added a 1948 John Deere Model B to the farm.

After retiring, Bob began writing a book on his favorite tractor, the Ford. The book was published in 1990, teaming Bob with renowned English automotive photographer Andrew Morland. Since then, Bob and Andrew have collaborated on ten books about classic tractors, and Bob has also authored five other tractor titles on his own.

Bob and his wife, Janice, now live in northern Wisconsin, almost within sight of the original homestead. Besides steady work on books, Bob and one of his three sons make about 150 gallons of maple syrup each spring.

About the Photographer

Andrew Morland was educated in Great Britain. He completed one year at Taunton College of Art in Somerset and then three years at London College of Printing studying photography. He has worked since graduation as a freelance photojournalist, traveling throughout Europe and North America. His work has been published in numerous magazines and books; related book titles include *Classic American Farm Tractors, Ford Tractors,* and *Fordson Tractors.* His interests include tractors, machinery, old motorbikes, and cars. He lives in a thatched cottage in Somerset, Great Britain, that was built in the 1680s. He is married and has one daughter.